AF553434

MOLECULAR FARMING

MOLECULAR FARMING

By
Dr. Amita Sarkar
Dept. of Zoology
Agra College
Agra (U.P.)
(India)

DISCOVERY PUBLISHING HOUSE PVT. LTD.
NEW DELHI-110 002

First Published-2009

ISBN 978-81-8356-420-5

Published by:
DISCOVERY PUBLISHING HOUSE PVT. LTD.
4831/24, Ansari Road, Prahlad Street,
Darya Ganj, New Delhi-110002 (India)
Phone: 23279245 • Fax: 91-11-23253475
E-mail: dphbooks@rediffmail.com
dphtemp@indiatimes.com
Website: www.discoverypublishinghouse.com

Printed at:
Sachin Printers
Delhi

Preface

The present title *"Molecular Farming"* has been written for undergraduate and post-graduate students of all Indian Universities. Molecular farming is basically the production of plant made pharmaceuticals and technical proteins. The present text deals with the utilization of plants as a source of raw materials and medicines. Form the earliest stages of civilization, plant extracts have been used to obtain technical materials and drugs to ease suffering and cure diseases. Mankind has been using this for thousand of years. Molecular farming has the potential to provide virtually unlimited quantities of recombinant antibodies, vaccines, blood substitutes, growth factors, cytokines, chemokines, and enzymes for use as diagnostic and therapeutic tools in health care, the life science and the chemical industry. The objective of this title is to harness the power of agriculture to cultivate and harvest plants or plant cells producing recombinant therapeutic, diagnostic industrial enzymes and green chemicals.

To make the work more comprehensive and informative, the author has consulted many authoritative books, research journals, abstracts, monographs etc., so there can be no claim to originality except in the manner of treatment.

The author expresses his thanks to his friends and colleagues whose continue inspirations have initiated him to bring out this book.

The author expresses his gratitude to Mr. Wasan and staff of M/s Discovery Publishing House Pvt. Ltd. for their whole hearted cooperation in the publication of this book.

In the mean time, the author will remain sincerely responsible for any shortcomings of the book and be grateful to the readers for their suggestions and constructive criticism for the continuous betterment of the book. He takes this opportunity to appeal to the readers to send their suggestions straightaway to his Publisher.

Author

CONTENTS

1

INTRODUCTION

The scientific revolution in drug discovery and product development that occurred at the end of the 20th century, i.e., the advent and full realization of biotechnology, continues unabated into the 21st century. The growth has been remarkable in technologies, products, companies, and profits for biotechnology. From its early era of product approvals numbering 60 in the 1980s and 1990s (an 18-year period), this discipline has increased the discovery, development, production, and commercialization of innovative biological products with about 80 more by 2006 (a six-year period). Moreover, now over 100 human disease conditions are treated with biotechnology products (an increase of 100%), many of which are still major medical break-throughs. The expansion of companies engaged in the development and marketing of biotech products also grew dramatically from 30 in 1999 to over 70 in 2006. Additionally in 2006, 200 companies have about 500 biological products in clinical research in patients.

Biotechnology is the use of a living system(s) to discover new disease biological targets and produce a pharmaceutically useful product, often called a biological product because of its structural and mechanistic similarity to naturally occurring substances in the human body. Biological products, developed and marketed, have been predominantly proteins of eight uniquely different types, including 8 *growth factors* (GFs), 3 blood factors, 30 hormones, 12 interferons, 18 monoclonal antibodies (Mabs), 15 enzymes, 6 fusion proteins, and 3 interleukins. New biologicals have expanded to include DNA/RNA (deoxyribonucleic acid/ ribonucleic acid) derivatives, such as 1 m-RNA (messenger-RNA) analogue, 4 peptides, 14 tissue or cell therapies, biologic drug carriers, such as 5 liposomes, and 16 natural derivatives.

Biotechnology further encompasses biological products that have agricultural uses and industrial applications. Farming is being revolutionized, such that genetically modified organisms in plants are being used in hundreds of millions of acres of food crops with manifold farming and public benefits, e.g., greater crop yields per acre, less insecticide and herbicide use for an improved environmental impact, and less cost per acre for farming. In industry, naturally occurring microorganisms are being studied and used to consume substances harmful to the environment, such as hydrocarbons (oil), mercury, and sulfuric acid. Biodegradable enzymes are used in manufacturing to replace toxic substances that use to enter the biosphere.

Biotechnology is a collection of biologic techniques and drug development technologies that permit whole new biologic discoveries and products. A plethora of techniques now exist starting with the three corner-stone technologies of *recombinant DNA* (r-DNA) technology, Mabs, and p*olymerase chain reactions* (PCRs). In product development, technologies also include genetics-related technologies such as, antisense, genomics, gene therapy, pharmacogenomics, ribozymes, and transgenic animals. Further processes are combinatorial chemistry, *high-throughput screening* (HTS), bioinformatics, proteomics, X-ray crystallography, receptorology and *protein kinases* (PKs), virology, and cell and tissue therapies. A major technological advance in biological product development over the last 10 years is molecular engineering,

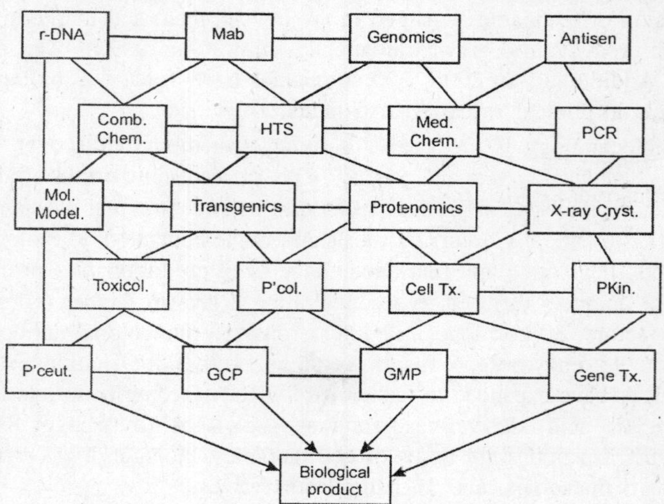

Fig. 1.1. Research and development network.

where in the biological molecule is manipulated regarding its amino acid or carbohydrate content to enhance biological function or reduce toxicity. All of these technologies and processes are collectively employed by the biotechnology industry to discover, develop, and produce biological products for patients.

These technologies help to elucidate new biologic mechanisms of disease, identify naturally occurring substances or processes responsible for a biologic effect, create duplicates of the natural substances that often are found only in minute amounts in the body, innovate new products that enhance natural processes against disease, block the function of dysfunctional proteins or nucleic acids, reduce action of natural processes gone awry as in inflammation in arthritis, and permit mass production of these rare products for commercialization. More than 140 biological products comprise a new "biological" method to treat human disease, i.e., biotherapy, an armamentarium for health-care providers, to be used in conjunction with drugs, devices, radiology, physical therapy, and psychotherapy.

History of Biotechnology

Some science historians will cite that the origins of biotechnology go back 4000–8000 years to the Sumerian, Egyptian, and Chinese cultures. Fermentation is an age-old, basic biologic process whereby a living organism, a yeast, will react with carbohydrate materials, such as wheat, in a vessel to produce alcohol. This Sumerian product was beer. This basic biotechnology process (fermentation) was employed also in the preparation of bread and cheese, food staples, and later wine over the millennia.

The modern era of biotechnology is thought to have started in the 1950s, with the discovery by Watson and Crick of the three-dimensional (3D) construct of the DNA double helix—the matched pairs of four nucleic acids (adenine, guanine, cytosine, and thymidine) in a specific sequence, parallel chains, and a 3D spatial configuration. Several key discoveries in biology in the 1960s underpin biotechnology, namely, the genetic code is universal in nature among all living things; for the 20 amino acids, 64 specific nucleic acid triplet codes are responsible for interpretation of genes into proteins, and genetic material is transferable among different organisms. The basic tenet of molecular biology is that DNA makes RNA through the process of transcription in the cell's nucleus, and RNA makes proteins, through the process of translation in the ribosome structures in the cell cytoplasm. Genetics began with the discoveries of Charles Darwin and Gregor Mendel in the mid-1800s; the principles were used in breeding animals and plants to enhance desirable traits. Proteins were discovered to exist in 1800s, and DNA being responsible for carrying genetic information was proved in 1940s.

The early 1 970s saw the development of the two core technologies of biotechnology, i.e., r-DNA and Mabs, which account for 80 of the 140 commercially available products in 2006. This r-DNA process is sometimes called "*genetic engineering.*" The r-DNA technology is a serial process, whereby (i) a protein is associated with a biologic action and is identified; (ii) a specific human gene (a DNA sequence) is associated with the protein; (iii) the human gene is inserted into bacterial DNA (plasmid); (iv) the plasmid is placed into the non-human host cells (e.g., *Escherichia coli* bacteria); and (v) the host cells manufacture their typical variety of proteins and produce a human protein from the human gene. Mabs are proteins that are produced by the body's plasma cells in response to foreign substances, and Mabs serve to attach to and neutralize foreign substances. Hybridoma technology was developed to produce Mabs for commercialization. Several discoveries have enhanced the r-DNA and Mab processes including PCR where genetic material can be accessed and reproduced even a million-fold; sophisticated analytical processes for better genes and proteins assessment; and accelerated gene-sequencing techniques that resulted in full description of the human genome as well as other species. The first commercial product derived from biotechnology was recombinant human insulin in 1982.

Biotechnology has become a major and common platform for new products approved for clinical use. For example, from 1998 to 2003, biotech research and development was responsible for 36% of all new molecular entities and all drug approvals by the Food and Drug Administration in U.S.A. About 140 biological products are approved for use in U.S.A. for over 100 indications.

Core Drug Development Technologies

The products produced through biotechnology, also called *biological products*, have been predominately proteins, which have been identified in the human body for their beneficial or malevolent actions, and then created ex vivo to be used later as a therapeutic to interact with human physiologic systems. About 100 of the 140 commercially available products are proteins. Proteins are complex, large molecules with several key structural features required for their activity. Each protein has a specific amino acid sequence from the 20 amino acids in nature, which must be preserved intact. Special structural features include disulfide bridges, special peptide domains often engendering functions, specific terminal amino acid species, glycosylation (carbohydrate species attached to the protein backbone), isoforms with naturally occurring protein variations in structure, and 3D configurations necessary for function, especially receptor interactions.

r-DNA Technology

The predominant technology in biotechnology from the 1980s to the present remains r-DNA technology, which has been the process used to create 60 plus products, that is, over 40% of the commercially available products. The basic notion is the manufacture of human proteins in non-human living systems, a form of genetic engineering. Five steps comprise r-DNA technology: protein identification, gene isolation, cloning of genetic material and expression of proteins, manufacturing (scale-up processes), and quality assurance for both protein and process integrity. Step 1 involves finding a protein responsible for some biological effect in the human body that has therapeutic potential. The protein needs to be isolated from its normal milieu, usually a body fluid or cell. The structure of the protein and its functions are determined, including all the structural features noted above.

Step 2 requires the isolation of the human gene responsible for the protein, which entails one of three mechanisms. First, we often will know the protein's full amino acid sequence, and we do know the 64 nucleic acid triplets that code for the 20 amino acids. Hence, we can construct many combinations of those triplet codes that may be genetic representations of the target gene for the target protein. These genetic constructs are genes, one of which will be identified through screening as the correct gene with the capacity to produce the target protein. Second, we may be able to find the human cell that produces our target protein. In this cell, we can ferret out the m-RNA that is responsible through the process of translation for producing the target protein. The viral enzyme, reverse transcriptase, is capable of creating the target complementary DNA, or gene, from this specific m-RNA. This method was used to find the gene for insulin, which led to marketed products. Third, the human gene could be fished out of the human genome using nucleic acid probes, which is a complex, daunting task. In this method, we identify several peptides in the amino acid sequence of the protein. Using the triplet codes for the amino acids in the peptide subunits of the protein, we build specific nucleic acid combinations (probes) for each peptide. Then, we break the chromosomes

Table 1.1. Recombinant DNA technology

Step 1	Protein work-up: Protein identification Protein isolation Biological property description Protein structure work-up – amino acid sequencing and mapping, glycosylation, disulfide bridging, peptide domains
Step 2	Gene isolation (three alternative methods): Nucleic acid triplet sequencing for amino acid sequence mRNA isolation, with reverse transcriptase and complementary DNA DNA probes from genetic library
Step 3	Cloning and expression: Gene (human) insertion into bacterial plasmids Plasmid incorporation into host cells Host cell production of proteins Master working cell bank created
Step 4	Manufacturing scale-up: Inoculum stage Fermentation or cell culture stage Protein purification stage Formulation stage
Step 5	Quality assurance: Genetic testing Bulk product testing Process validation Final product testing

(3–4 billion pairs of nucleic acids) into thousands of pieces of DNA. Through many serial experiments, we try to match the first nucleic acid probe to the DNA mixture, which does create a subset of matching DNA pieces (hundreds or thousands). In another series of experiments, a second, different nucleic acid probe for a different peptide is matched against this DNA subset for further matches. Matches do occur, resulting in a series of possible genes. Each gene must be evaluated by genetic analyses to insure production of the correct protein, which then are tested to be sure that the protein has the structural features and pharmacological properties in test animals of the targeted, naturally occurring protein. The gene for the protein, epoietin alfa for anemia, was discovered by this laborious method.

Step 3 in r-DNA technology involves cloning of the gene and expression of the protein by the gene. Cloning is the reproduction of the target human gene in a non-human cell. Expression is the production of the target human protein by a nonhuman cell containing the human gene. These processes require a vector for the DNA (genes), so that the gene can be carried into a host cell. A bacterial plasmid is a circular piece of DNA that is transferable between cells (therefore, a carrier), will accept the insertion of a human gene, and will allow the human gene to be turned on. Plasmids are further manipulated to maximize their function with the addition of promoter, enhancer, and operator DNA sequences. The plasmid must be cut open to accept the human DNA (gene) by unique enzymes (bacterial restriction endonucleases), each of which is highly specific to a certain nucleic acid sequence that can match a terminal end of the human gene. These "sticky" ends of the opened bacterial plasmid and the human gene permit recombination of the DNA, under the influence of a ligase enzyme, resulting in an r-DNA molecule containing a human gene inserted into a bacterial plasmid. Next, the r-DNA molecule is inserted into a host cell, which serves to produce all of its routine proteins, and the cell manufactures the human protein from the human gene that it carries. The host cell needs to possess a set of demanding characteristics to be used feasibly and cost-effectively in manufacturing processes: a short reproductive life cycle, long-term viability in an in vitro setting, the ability to accept bacterial plasmids, substantial productive capacity (yield) for proteins, the ability to produce the human protein consistently without its alteration, possibly glycosylation of the protein if needed for its activity, ease of manufacturing in the later scale-up process, as low as possible cost in manufacturing, and patentability to protect the intellectual property. The host cells can be bacteria, usually *E. coli*, yeast cells, and mammalian cells, usually Chinese hamster ovary cells or baby kidney hamster cells. This unique newly created cell and its offspring, created in the laboratory, are called the master working cell bank.

Step 4 in r-DNA technology is scale-up manufacturing and comprises four phases: inoculum, fermentation, purification, and formulation. Inoculum phase entails the use of daughter cells from the new host cell (master working cell bank), actually removed from its storage in a –70°C freezer. The daughter cells are grown in specific media in serially larger flasks and assessed for normal growth characteristics. The growth medium (liquid and air) is a unique and specific mixture of minerals, compounds, and nutrients to enhance cell viability (lifespan) in vitro and functional ability of cells to produce proteins (maximize yield). The fermentation or cell culture phase involves inoculating thousand of containers, or large fermenters, with cells from the inoculum phase and adding the appropriate fortified growth media. The host cells will proceed to produce proteins, either intracellularly in storage vacuoles for most bacterial host cells or extracellularly into the media for mammalian cells. Feeding of the host cells and removing waste from the media need to be done periodically to sustain host viability and productivity. Harvesting the cells and media is the next step to obtain the protein. Purification of the protein follows, which varies between bacterial and mammalian systems. For bacteria, the cells are removed from the liquid in the fermenters by centrifugation into a cell paste, which is centrifuged again to break out proteins from the cells. The protein mixture is then run through an extraction process, often *high-pressure liquid chromatography* (HPLC) to separate the target protein from all other proteins. For mammalian cells, the culture media contains the proteins, which were secreted extracellularly by the mammalian cells, and is collected periodically. Extraction and purification are basically similar processes for the mammalian process. A pure bulk protein is the result of purification.

The final phase is formulation, wherein a diluent is chosen for the protein, incorporating the best mix of fluids, buffers, stabilizers, and minerals to achieve optimal protein stability, maximal shelf life, and patient acceptability. Sterile water, normal saline, and dextrose 5% in water are three common diluents. Variables to deal with include protein traits and patient-disease factors. Formulation of proteins is confounded by the general delicate nature of proteins and the many degrading processes that can occur with them.

Step 5 in r-DNA technology is quality control for the final product, components, and processes throughout the manufacturing process. As can be readily observed from this description of manufacturing by r-DNA biotechnology, manufacturing is quite multifaceted and complex with much potential for contamination, variation, and alteration, leading to poor outcomes, e.g., a deteriorated product, a degraded protein, viral or bacterial contamination, poor yield, an immune reaction in

Table 1.2. Biological product—stability/degradation

Precipitation
Clumping/aggregation
Cross-linkage
Unlinkage (disulfide bridges)
Amino acid mutation
Glycosylation/deglycosylation
Conjugation
Amino acid deletions/additions
Reduction
Oxidation
Folding/unfolding of protein
Deamidation
Proteolysis
Protein inclusions
Terminal amino acids variations

patients, and even a different protein. Contamination can result from typical methods for drugs, e.g., microbial or chemical means, and, because of the genetic material employed, through oncogenic or viral DNA incorporation or alteration. Therefore, quality control ensures final product integrity through an extensive series of tests, which involve four key areas: genetic material (plasmids and genes), bulk protein product, final product, and the manufacturing process.

Mab Production

Mabs are complex proteins that have a uniform basic structure, comprising four subunits that are divided into two matched pairs of protein material—two heavy chains and two light chains, linked by disulfide bridges, forming a "Y" configuration. One end, the variable region with light chains, binds to the target antigen [*complement determining region* (CDR)], while the other end of the molecule, the constant region with the heavy chains, initiates an immune reaction. These Mabs are highly specific proteins that the plasma cells in the human produce, a single Mab, against a single antigenic foreign material. Historically, Mabs have been produced in biotechnology in a mouse, and are targeted against human antigens. We are developing a product (Mab) to attack and eliminate these target antigens. The mouse will create highly specific Mabs against the human antigen (target); however, the quantity of Mabs (protein) produced by even very large numbers of

Table 1.3. Quality control tests in r-DNA technology

Genetic material: (5–10 tests)
- Karyotypic analysis
- Oncogene screening Gene stability
- Infection DNA screens

Bulk protein products: (20 tests)
- Amino acid sequence
- Peptide maps
- HPLC
- Radioimmunoassay
- Western blot chromatography
- Bioassay

Process validation: (10 tests)
- Protein yield
- Protein challenge
- Endotoxin spiking

Final product: (30 tests)
- Protein analyses (repeat of #2 above)
- DNA contamination
- Stability tests
- Freeze–thaw tests

mice is very small. Hence, biotechnology for Mabs also uses a myeloma cell and fuses it with the mouse plasma cell to create a murine hybridoma cell. Myeloma cells impart several additional characteristics to the hybridoma, i.e., very high production of proteins (Mabs) and long lifespan, which make Mab manufacturing commercially feasible. These hydridoma cells are the new master working cells to produce large amounts of specifically targeted Mabs.

The murine origin of Mabs creates significant limits for these products, because administration of murine Mabs to humans leads to the body's rejection against the murine nature of the protein, an immune reaction. Two limits of murine Mabs are a toxic *human antibody mouse antigen* (HAMA) response with fever and chills and less binding of the Mabs to the target cell, thus limiting their activity. Immunogenicity is a complex subject with Mabs, and a human can react with various types of Mabs against the Mab being administered, which will neutralize, clear, sustain, cause HAMA, cause allergic reactions, or do nothing. A patient's genetics and current immune status will impact immunogenicity as well as the protein and its formulation. Therefore, scientists now manipulate Mabs by substituting human subunits for the four murine subunits, creating chimeric molecules, part murine (about 25%), and part human (about 75%). Even more humanization can be achieved to about 90%, called "*humanized*" Mabs, with only the CDRs (identifying and binding to antigens) being murine. Furthermore, mice have been bred by genetic engineering to produce fully human antibodies. This humanization lessens toxicity and can result in an increased activity. Because of the more human Mabs, we now have been able to expand the number (18 Mab products) and uses of Mabs to manage inflammatory conditions, treat cancer, or bind to proteins to arrest a process, as in slowing kidney rejection in transplant patients. Yet another more recent development in antibody discovery is the use of phage displays instead of hybridoma technology to create the Mab.

Biological Techniques and Processes

Most of the biological products have been created through protein r-DNA and Mab technologies (over 80%), as stated before. However, biotechnology has expanded in its technologies over the last 20 years to encompass a breadth of areas.

Table 1.4. Biotechnology techniques and processes

Polymerase chain reaction	Bioinformatics
Recombinant DNA proteins (proteins)	Molecular engineering
Monoclonal antibodies	(AA/CHO/PEG/Fusion)
Structure/activity relationship	Peptides
Genomics	Receptors
Gene therapy	Animal-based products
Ribozymes	Marine-based products
Nucleotide blockade	Tissue engineering
Pharmacogenomics	Cell therapy
Transgenic animals	Virology
Combinatorial chemistry	Formulations
High-throughput screening	Liposomes/Polymers
Protein kinases	Biosimilars (Biogenerics)
Proteomics	

PCR

PCR is a critical core process in biotechnology, which permits the expansion of the amount of genetic material (DNA), starting from minute amounts. The process involves first denaturing DNA with high heat (90°C), i.e., unraveling the DNA double helix so that the genetic code (sequence) can be read and possibly duplicated. Second, a leader sequence for DNA is used to bind to the target DNA and initiate reading of the genetic code at a specific point. Both helices and strands of DNA can be read, that is, duplication of the target DNA sequence. Third, the heat-stable enzyme from the *Thermus aquatis* bacteria, DNA polymerase, catalyses the reading of the genetic code with extension of the replicated DNA sequence. By sequential repetition of these three steps, the genetic material is magnified; for example, 20 replications yield a million-fold increase in the DNA material.

Genetic Technologies

In drug discovery and development, genetic materials (DNA, m-RNA, genes, and RNA enzymes) have come to the forefront as potential biological products and, more so, as primary tools of product discovery. Six primary areas are covered here: antisense, gene therapy, genomics, pharmacogenomics, ribozymes, and transgenic animals. Antisense is a RNA molecule that is complementary to, or a mirror image of, a segment of an aberrant or mutated m-RNA molecule, involved in the pathogenesis of a disease. The antisense RNA molecule will bind to the noxious m-RNA molecule, preventing the disease from manifesting. One product is currently available, fornivirsen, used to treat cytomegalovirus associated retinitis that can occur in AIDS patients.

Gene therapy is a technology employing a gene as a therapeutic agent to treat a disease. The potential goals of gene therapy include replacement of an inactive gene, reactivating inactive genes, turning off genes causing disease, as in oncogenes, turning on further naturally protective genes, or adding a gene characteristic to cells, such as increased susceptibility of cancer cells to chemotherapy drugs. Development challenges in gene therapy include finding the target gene that is causing the disease or needing enhancements, identifying the optimal target human cell for the disease and patient for insertion of an additional gene, inserting (delivery) an extra gene reliably into human cells *ex vivo* or *in vivo*, and causing the gene to be functional in a reasonably physiologic manner (extent and duration of activity). Science is rapidly isolating and identifying all the genes in the human genome. However, gene delivery has not been achieved in a reliable, reproducible manner sufficient for routine therapy.

Viruses are most commonly used for gene delivery with the investigational therapies because of their natural ability to carry genetic material, deliver it into human cells (infect or transfect), and allow the genes to be turned on. However, the optimal gene vector has a lengthy and daunting list of optimal traits, e.g., packaging of the DNA of varied sizes, protection of the DNA package in vials and in vivo, serum stability with nucleases, targeting to a specific cell type, infection of non-dividing target cells, internalization of vector/gene by cells, endolysosomal protection in cells, endolysosomal escape, cytoplasmic transport, nuclear localization and transport, efficient unpackaging in nucleus, gene activation with transcription and then translation, stability of the gene expression, and robustness of gene expression. Additionally, clinical use of a gene/vector product in patients needs ease of administration and safety in patients, that is, little toxicity, no immunogenicity, and non-pathogenesis on its own. Manufacturing needs for a vector/gene package include ease of fabrication, inexpensive synthesis, and facile purification.

The study of genomics involves the identification of the sequence (genotype), the activities of all the 30,000 genes in the human body, and their expression into proteins in specific patient groups (phenotype). Genomics is a technology in which we search for a gene that is responsible for some process or product in the human body. The gene may lead to a new process or disease target for which drugs or biologicals could be developed to favorably alter. For example, the gene for cell immortality was discovered, which led to the protein, telomerase, an enzyme. Telomerase is responsible for adding telomeres, short-nucleic acid sequences, to the end of all chromosomes, protecting chromosomes from mutations and, ultimately, cell death. Alternatively, genomics may lead to a gene that produces a protein, which can be used as a therapeutic agent. For example, osteoprotogerin is a key protein that is the natural substance that turns off osteoclasts in bones, which break down bones and could lead to osteoporosis. Genomics requires the collection of massive amounts of information regarding the human genome or genetics, including protein and peptides, receptors for activity of proteins or drugs, mechanisms of drug activity, and subunits of drugs, proteins, RNA, or DNA responsible for physiologic or pharmacologic actions. The science of bioinformatics now exists to store, share, integrate, analyze, and manipulate these massive amounts of data via facilitated methods using computers and algorithms to identify new products.

Ribozymes are molecules comprising sequences of nucleic acids that possess enzymatic catalytic properties to bind to specific sites in

DNA or RNA and cleave the chain. A ribozyme will have subunits responsible for the binding function and subunits responsible for enzyme function. They generally have the following desirable traits; specificity in targeting, cleavage of RNA, small size amenable to formulation and dosing, and multiple turnover (one molecule binds and acts and then moves on to next molecule and repeats its function). However, challenges include need for cell insertion (transfection), nuclease protection in the blood, and chaperone proteins for movement in cell cytoplasm. Inhibitory RNA molecules, known as *antisense* and siRNA, are being studied because of their excellent specificity and function to turn off aberrant RNA, but the disadvantages of ribozymes also exist for antisense, plus they require a vector as in gene therapy for cell entry.

Pharmacogenomics is the study of genetic phenotypes of patient groups and their impact on drug actions, changing drug pharmacokinetics and/or activity (more or less pharmacologic effects). Over 60,000 *single nucleotide polymorphisms* (SNPs) exist in exons that are genetic changes, deletions, insertions, or repeats. For drug activity, these SNPs can change action of metabolic enzymes, receptor activity, or drug transport, any of which can lead to more adverse events, dosing changes up or down to achieve the same effect, and disease subtypes with different drug responses. Hopefully, we can identify (diagnose) these abnormal or just different genetic phenotypes responsible for different disease presentation or drug actions, and either use a drug in a more effective subpopulation, or change dosing for less side effects or more drug efficacy. For example in oncology, some breast cancer patients will have a genetic abnormality being Her2-Neu gene-positive, leading to more aggressive and fatal disease course. Fortunately, a Mab has been developed against this genetic subpopulation, trastuzumab (Herceptin), offering more efficacy only in these patients. Viability of pharmacogenomics in patient care requires diagnostic tests to be identified, validated, and commercialized to identify these patients at risk, and then requires the studies to establish appropriate drug doses for the unique phenotypes or drugs that will work in these unique phenotypes.

Transgenic animals are another genetics-related development in biotechnology and actually all drug development that enhances the screening of potential therapeutic molecules. Through genetic engineering, an animal's (most often mice and rats) gene makeup is altered by knocking out genes or gene additions, which results in the animal presenting with a disease that is more human-like in its pathology. A potential new drug candidate is administered to these transgenic animals, and the animal will present disease models that will better predict (respond to) the drug's action as being representative of what would really occur in humans. Another use of transgenic animals is for the manufacture of biological products in the animal. Through genetic engineering, human genes are added to the animal make-up often within the mammary structure of usually sheep, goats, or cattle, such that the human gene produces a human protein in the animal's milk, which can be harvested and the human protein separated.

HTS

HTS is intended to obtain faster more high-quality product leads from large volumes of genetic or peptide molecules. HTS has improved the number of molecules that can be screened for activity by 10- to 1000-fold. The process of HTS is dependent on improved analytical processes (better surface chemistry, capture agents, and detection methods), miniaturization of equipment, and automation. Currently, over 100,000 samples can be tested in a day.

Combinatorial Chemistry

Combinatorial chemistry involves the use of the basic building blocks in biochemistry, either the 20 amino acids or 4 nucleic acids, to build new molecules. All the different combinations of a set number of building blocks can be created; for example, the use of 10 different amino acids can result in over 3.5 million decapeptide compounds. Huge libraries of compounds are produced, which require screening through HTS and the use of informatics to help sort out the structures and especially the activities of all these new compounds.

Proteomics

The proteome is the complete protein makeup in the human body. Proteomics is the study of protein structures and their properties. The proteome is more complex than the genome when we consider the greater complexity of proteins, e.g., 20 amino acids vs. 4 nucleic acids, and their manifold structural requirements, including the amino acid sequence, disulfide bridges, glycosylation of proteins, the complex carbohydrate structures, the amino- and carboxyl ends of proteins and their variation, the isoforms of the same protein in one patient and between patients, and the 3D configuration of proteins. Proteins have a certain mass, isoelectric point, and hydrophobicity impacting their activity. Protein function will also potentially change during development of the fetus and child, in disease vs. normal physiology, during inflammation vs. none, and possible different actions at specific tissue sites. All of these different properties will require sophisticated and sensitive analytical technologies to identify and understand protein structure and function. Proteomics assists us in finding new disease targets and possible biological products for therapy.

Signal Transduction and PKs

Over the last ten and, especially, five years, cell function is being more fully elucidated, and a group of protein enzymes, PKs, has been found to play principal roles in the communication between and within all cells resulting in activation of all cells. Several thousand PKs exist and are very specific to certain cells and cell functions. Aberrant or excessive cell activity can be mediated by the PKs, contributing to diseases such as cancers or inflammatory conditions. PKS become targets for drug intervention to turn off or reduce their activity and moderate a disease. They are quite desirable targets given their universality in cells, very high specificity to a cell function, involvement in many principle cell functions, their specific protein structures, the identifiable mechanisms of action within their substructures, and their accessible sites for drugs and biologicals. For example, tyrosine PKs serve as cell receptors and allow binding to a cell of various cell-to-cell communication ligands such as hormones or GFs that leads to autophosphorylation and activation of the PKS. Their structure includes three main components (domains), one extracellular that binds the ligand, a segment in the cell wall, and then an intracellular segment that, after ligand binding, initiates a series of cell actions, activating other intracellular proteins that eventually result in cell action or production of cell proteins to yield downstream actions. The drug, imitanib (Gleevec), was the first PK inhibitor approved for use and treats chronic myelogenous leukemia with a bcr-abl SNP change that creates the Philadelphia chromosome positive abnormality. Imitanib disrupts the function of the intracellular domain to turn off the abnormally functioning cell.

Cell and Tissue Therapy

Another technique to treat disease is based on obtaining healthy cells from a specific tissue, selecting out a specific subset of cells with certain desirable properties, and enhancing the activity of these cells through ex vivo manipulation. We then return these specifically selected, enhanced, and activated cells to patients whose cells are not sufficiently functional, thereby ameliorating a disease. Currently, chrondrocytes responsible for cartilage production are taken from a patient's knee that has serious damage and is repairing poorly. These chrondocyctes are manipulated *ex vivo* and returned to the patient to normalize cartilage production. Bone marrow progenitor cells are collected from peripheral blood, bone marrow cells, or cord blood, and the cells with greatest regenerative potential are selected through various cell-tagging processes. Following life-threatening chemotherapy in a cancer patient, which destroys almost all the bone marrow, these selected progenitor cells are administered to the patient to accelerate regeneration of bone marrow and white blood cell production, thereby preventing infections. Foreskins of newborns are collected and placed in a 3D construct that permits growth of dermis and epidermis. This construct is used later in venous leg ulcers to accelerate wound healing, thereby reducing health-care needs. In tissue engineering, generally, we need tissue from patients or normal donors via biopsy, a process for ex vivo cell expansion, a scaffold on which to grow the cells, and a bioreactor system, in which to grow the new tissue into its full size, normal structure, functional capacity, and devoid of toxicity, immunogenicity, or fibrosis.

Molecular Manipulation of Biologicals

A new generation of biological products is the derivative of existing molecules that are altered to change their properties resulting in new molecules. Molecular manipulation of biologicals is now a major technique used in product development, and has been expanded to include amino acid changes (truncation of domains or replacement of amino acids), glycosylation changes, pegylation, fusion of proteins and drugs, isoform selection, monomer vs. dimer molecule selection, and humanization with antibodies. The properties of proteins that can be altered include receptor affinity, receptor selectivity, pharmacokinetics (half-life, metabolism), immunogenicity or antigenicity, effector function, potency, safety (adverse events), stability, solubility, absorption by new routes of administration, and manufacturing yield. Besides enhanced properties, these new biological molecules have the added benefit of being patentable as new drugs.

Protein manipulation for enzyme proteins can serve as a good example of several possible manipulations. Alteplase is a thrombolysis enzyme protein used to prevent death in acute myocardial infarction. The protein has 527 amino acids, 5 disulfide bridges, glycosylation at several amino acids, and 5 peptide domains. Truncation of most domains leaving intact the protease domain resulted in a new biological, reteplase

(Retevase), with quite similar thrombolytic activity. Another set of changes was made to the alteplase protein, that is, six amino acid exchanges at three sites, along with added glycosylation at one site, resulted in a new protein, tenecteplase (TNKase). This new protein has improved administration, intravenous bolus instead of infusion, more fibrin specificity, less degradative enzyme susceptibility, no added toxicity or immunogenicity, and yet the pharmacologic action being sustained.

Pegylation is a process in which polyethylene glycol (PEG) is added to the protein structure, and it has been done for alpha-interferons (2a and 2b), a growth factor (filgrastim), a liposome (doxorubicin), and enzymes (asparaginase and ademase). Pegylation will vary with the number and types of PEG molecules, the attachment site on the protein, and the linker molecule for the PEG to the protein. Any of these will alter the pharmacokinetics or activity of the protein. The addition of PEG may extend the product's half-life or duration of effect, as in alpha-interferon-2a (Pegasys), permitting less frequent dosing, yet the desired activity of the molecule can be maintained. Glycosylation entails changing or adding carbohydrate species, e.g., sialic acid residues, to the amino acid backbone of a protein. The altered carbohydrate structure (hyperglycosylation) in the protein, epoetin alfa, created the biological, aranesp; the new product extended its half-life about threefold, again offering less frequent dosing, while sustaining its pharmacologic properties. Altering glycosylation may require changes in amino acid sequence, because only certain amino acids carry the carbohydrate structures, asparagine, serine, and threonine. Also, the type of sugars and their structures are variable and affect the action of the biological (e.g., mannose, sialic acid, galactose, fucose, and *n*-acetylcysteine).

The concept with fusion proteins is to combine two compounds, two proteins, or a protein and a chemical. Goals have been to alter pharmacokinetics, result in combined properties for the full molecule, serve as a carrier to the site of action, or enhance absorption of the large biological molecules. Several examples have been achieved and marketed, e.g., gemtuzumab, a Mab plus a cytotoxic cancer drug for acute myeloid leukemia; denileukin, a protein (IL-2) and a toxin, diphtheria toxin for renal cell carcinoma; alefacept, a CD-binding protein and a fragment of immunoglobulin-G1 (IgG1) for psoriasis; abatacept, the CTL4-A cell-binding protein and a fragment of IgG1 for rheumatoid arthritis.

Engineering of Mabs to serve as carriers or fusion proteins involves consideration of the target antigen, the antibody-delivery vehicle, and the linker technology. The target antigen (Ag) often is a cell surface protein serving as a cell receptor for cell activation, also called "CD" complement determining region; any one cell can have hundreds of CDs on its surface. The Ag for a Mab should be specific to the desired antibody (Ab) with high-binding affinity and expressed in the tumor or abnormal tissue at higher levels, homogeneous at the cell surface, in significant percentages of patients, expressed throughout the disease process, and not or minimally expressed in normal tissue. The antibody should possess accessible sites for loading the second chemical or product and be able to release the "*conjugate*" at the target tissue site. The conjugate being added should not alter Ag–Ab binding, Mab internalization into target cells, effector function of the Mab such as antibody-dependent cell cytotoxicity, pharmacokinetics of the Mab, biodistribution of the Mab conjugate, or the aggregation, immunogenicity, or toxicity of the Mab conjugate. The linkers for the Mab and conjugate should have several desirable properties as well, such as easy site-specific attachment, no toxicity, stability in plasma, cleaved intracellularly, and no immunogenicity. These many desirable properties listed above display the high challenge in developing these Mab conjugates. Mabs also are engineered as antibody fragments to be used alone or combined with other protein as fusion proteins for their combined actions. Abatacept (Orencia) combines the Fc fragment of IgG1 and the T-cell antigen protein, CTL4, which creates a protein molecule with specificity for T-lymphocytes to turn off the autoimmune reaction in rheumatoid arthritis.

Animal and Marine Products

Identification of pharmacologic activity of biological products in animals has been a process for a long time in drug discovery; of course, porcine and beef insulin were extracted from the pancreas of animals and used in patients for most of the 20th century. Then, in 1982, a recombinant form of human insulin was created and marketed. Animal proteins may be different in their construct from humans, yet present useful pharmacologic properties that can then be used in humans as therapeutic agents. We now have a variety of biological products isolated and identified from animals, mostly proteins, and then they have been reproduced by recombinant technologies and are used to treat human diseases. From leeches, we have two thrombolytic enzymes, bivalirudin and lepirudin. The Gila monster provides a protein for diabetes mellitus, exenatide. The marine snail is a source to identify a protein for severe pain, ziconatide. Snakes have given us two enzymes, tirobifan and eptifibatide, to be used as thrombolytics in cardiovascular conditions such as acute coronary syndromes and angioplasties.

Biogenerics or Biosimilars

The creation of generic biological products currently is controversial and fraught with difficult manufacturing and patient-care issues. The basic tenet to this date by drug regulators in U.S.A. and the European Union (E.U.) is "the same manufacturing process for biologicals is not the same at different companies at different locales;" however, guidelines are being drafted by the U.S.A. and E.U. to permit generic biological product manufacturing and marketing. Proteins are complex molecules in their sequencing, carbohydrate content, post-translational modifications, 3D configuration, and immunogenicity. The formulation can even impact the immunogenicity of the protein in patients. Biologicals are manufactured in genetically engineered living systems with the great potential for many different alterations, contaminations, or degradations in perhaps unpredictable ways. Host cells can vary in type, stability, growth, and production. Vectors for genes also vary in type, site of gene insertion, their genetic sequence, use of promoter or enhancer gene sequences, and activity. Culture conditions are particularly variable and can impact protein production as well. Then harvesting and purification are yet further variables. Also, product analysis is very complex because of the complexity of both the molecules themselves and the manufacturing processes, as described earlier. Generic biological products will be available in the future. Some current justification for generic biological production is the example of growth hormones. There are now nine products marketed in the U.S.A., and their differences are relatively minor. However, each product was required to submit a full regulatory application for approval including clinical trials establishing safety and efficacy.

Biomarkers

The diagnosis of disease basically always has been dependent on various disease indicators related to normal human physiology and the pathology of the disease, such as liver enzymes, the kidney's urine electrolytes and proteins, and cardiac enzymes, to judge organ function or their disease status. These diagnostic factors are essentially disease biomarkers. Genetics has created added parameters in identifying patient-specific differences in disease occurrence or disease severity. Pharmacogenomics takes genetics one step farther in associating differences in drug activity in definable subpopulations of patients with specific genetic differences (inherited or acquired), especially drug metabolism. A drug or biological product that is active in this subpopulation with a genetic variation, and may only be effective in that subpopulation of patients, would require a biological disease indicator, a biomarker, to identify these different patients. Hence, another outgrowth of biotechnology is the biology of disease and the search for biomarkers to identify groups of patients with different responses to certain drugs, either drugs or biologicals. Therapy thus can be better individualized and given to the patient most likely to respond or avoid a predictable adverse event related to genetic differences. Several biological products are now effective and marketed for a certain disease and a specific subpopulation. For example, trastuzumab is available and only effective for breast cancer, given along with their chemotherapy, in patients with a very aggressive form of the cancer, who possess the oncogenic disease marker, Her2Neu. The PK inhibitor drug, imitanib, is only effective in chronic myelogenous leukemia with Philadelphia chromosome-positive status. The fusion protein Mab, gemtuzumab, is effective only in acute myelogenous leukemia with CD33-positive cells. These three exemplary treatments require a diagnostic test for the respective biomarker to document the presence of the genetic abnormality, engendering a higher likelihood of drug response. Biomarkers need to relate directly to the disease pathogenesis and the genetic or physiologic abnormality, be consistently present over the course of the disease, possess reliability to avoid false positives or negatives, must be validated in clinical trials, need to be practical in their conduct for anyone skilled in lab procedures to perform, and not be too expensive.

Nanobiotechnology

The nascent field of nanobiotechnology involves the use of and study of particles, organelles, instruments, drugs, and devices that measure or function within a size parameter of 1–100 nm. A nanometer is one billionth of a meter in size (10^{-9}). To give these sizes some relevance, the red blood cell size is about 5 μm (5000 nm), and an aspirin molecule is less than 1 nm. Biologic applications are at a very early stage of scientific evolution including the areas of bioanalysis, drug delivery, therapeutics, biosensors, and medical devices including tissue engineering; however, substantial financial investment from government, universities, and venture capital already exists. Drug discovery and diagnostics are the two most significant areas of research in the private sector. Currently, the scanning probe microscope is a tool in common use for cellular study, functioning on a nanometer scale. Nanoparticles can provide new labeling technology in cell or molecule analysis, based on changing color with changing particle size. Use as contrast agents in X-ray imaging is an application, possibly with better image resolution, tissue targeting, and retention in blood. Nanoparticles can be carriers for drugs in very minute amounts possibly enhancing drug penetration across membranes, changing drug solubilities, and altering pharmacokinetics. Liposomes can function as nanoparticles. The commercially available product Abraxane binds the cancer drug paclitaxel to albumin, which can be considered a nanoparticle biological drug formulation. In tissue engineering, artificial bone matrix in nanoparticle size is being studied.

Biological Product Delivery and Formulations

Most biological products (75%) are proteins, which, as described earlier, are large, complex, and delicate molecules. Formulations of proteins are influenced by protein traits, such as peptide lability, the 3D structure, the charge on the protein, and its stability. Breakdown of proteins can occur through many mechanisms (chemical and physical), also noted earlier. Further contamination and impurities are a significant potential problem, given the living system for manufacture. The final formulations often are sensitive to temperature extremes, which can cause aggregation or precipitation. Restrictions in diluents exist because of potential adverse changes in stability. For example, filgrastim growth factor requires dextrose 5% in water and not saline for the diluent, and sargramostim growth factor is the opposite. The protein formulations usually do not contain preservatives because of their interactions with proteins causing instability of the protein, which necessitates single-use vials. Refrigeration is the norm to achieve maximal shelf life. Prior to administration, vials should be warmed to room temperature to lessen local reactions. Some proteins will require lyophilization (a dry frozen powder instead of liquid) to create a practical shelf life. Excessive agitation is yet another possible denaturing problem for proteins. In creating the marketed product, the formulation for any biological molecule should avoid changes in activity of the molecule, its pharmacokinetics, its immunogenicity, its toxicity, and local irritation upon administration.

Biological Product Categories

Over 60 products (biologicals) were marketed by over 30 companies in U.S.A. from 1982 to 1999 (an 18-year period). In the last six years,

80 more products have been marketed. In many diseases, biological products often have been major breakthroughs offering the first treatments where nothing previously was effective for serious diseases, i.e., dornase for cystic fibrosis or beta-interferon for multiple sclerosis. Now, we have biological products in more than eleven distinct categories. The 157 products for the 123 distinct molecules are used for 121 separate indications and were created by 63 biotechnology companies. Four categories rank high in the number of commercial products, i.e., hormones, tissues and cells, antibodies, and enzymes. Following the discussion of each category of biological products, each table of biologicals approved for use in U.S.A. includes the generic and brand names, company marketing the product, and therapeutic areas of use.

Table 1.5. Marketed biological products

Types of products	*Molecules*	*Products*	*Indications*	*Companies*
Hormones	14	32	15	13
Antibodies (Mabs)	19	18	18	21
Enzymes	15	15	16	12
GFs	7	8	12	4
Interferons	10	12	10	8
Blood factors	4	4	4	5
ILs	3	3	4	4
Vaccines	5	6	4	3
Liposomes	5	5	5	5
Tissues and cells	9	19	19	20
Blood products; natural extracts	15	18	21	16
Other	17	17	18	18
Total	123	157	121	63

Hormones

Hormones comprise the largest class of biological products. Two naturally occurring protein-based hormones, insulin and growth hormone (somatotropin), have been duplicated through r-DNA technology into 20 products. Insulin was the first product from biotechnology to be approved in 1982, and 11 different types of products are now available with different salt forms or molecular arrangements, creating different activity (time-course) profiles. In Europe, six additional insulin products are available. One parent molecule can result in more than one product, because manipulations of the molecule can be done (for example, shifting terminal amino acids or use of monomers vs. dimers), without altering the biological purpose of the protein. Each molecule may have a related but different indication or pharmacokinetic profile. Nine *growth hormone* (GH) products are on the market for six indications. Fertility hormones are another large class of biologicals, six products in the U.S.A. and one in Europe (Puregon). The generic names of products often will be different for similar products produced by different companies, e.g., follitropin beta from Organon and follitropin alfa from Serono.

Interferons

These proteins called interferons are produced by many cells in the human body and participate in our immune system to provide protection against foreign substances, such an infectious material, and are involved in immune diseases. Interferons have several beneficial properties, an indirect mechanism of action, that is, stimulating the immune system, especially, the lymphocytes, and also possessing direct cytolytic and antiviral activity. Three families of interferons exist—alpha, gamma, and beta; and 11 products have been created. The indications for interferons are broad, including oncology (e.g., malignant melanoma and chronic myelogenous leukemia), viral infection (e.g., hepatitis C), and autoimmune disease (e.g., multiple sclerosis). Newer molecular forms include pegylated molecules that provide the advantages of longer duration of action and hence less frequent dosing, while sustaining the antiviral activity against Hepatitis C. Indications often are added to a product over time following the original product approval. Alpha-interferon is a very good example of this drug development process. The first indication was hairy cell leukemia in 1986, a narrow use for a severe oncologic problem, which was followed by approval for seven additional uses over the subsequent 10 years. Extensive clinical trials and a supplemental new drug application were required to establish the safety, efficacy, and appropriate dosing for each use. In Europe, six additional interferon products are marketed and even more in Asia. Hepatitis is a much more common disease in Asia and Europe.

GFs

These proteins can be divided into two areas: blood cell GFs, also known as *colony-stimulating factors* (CSFs), and tissue growth factors, all of which are ligands to communicate between cells. They are secreted by one organ's cells, e.g., erythropoietin by the kidney, and stimulate another cell type to produce an effect; in this example, bone marrow erythroid progenitors to accelerate their production of red blood cells and correct anemia. All these GFs are produced by r-

DNA technology. Twelve CSF products are available worldwide, i.e., filgrastim, pegfilgrastim, and sargramostim in U.S.A., and molgramostim, regramostim, lenograstim, and nartograstim in rest of the world, which stimulate white blood cell production and limit infectious complications in cancer patients. Two epoetin molecules (epoeitin alfa and epoeitin beta) stimulate red blood cell production, with two US products available, along with Eprex, NeoRecormon, and Epogin in Europe and Asia. Aranesp in U.S.A. and Nespo in Europe are the hyperglycosylated products of epoietin that have the extended half-life and less frequent dosing. Becaplermin is a tissue growth factor for the epidermis, being used to accelerate wound healing in diabetic ulcers. The second recombinant tissue growth factor is palifermin, impacting keratinocytes, and is used to more rapidly resolve the mucositis in cancer patients receiving chemotherapy, radiation therapy, and bone marrow transplants.

Blood Coagulation Factors and Interleukins

Blood factors are proteins involved in normal blood coagulation as cofactors in the coagulation cascade. The deficiency of any one blood factor leads to serious bleeding disorders (hemophilia), but it is fully correctable through replacement therapy with these r-DNA proteins. Factor 8 is available in seven products, all with the same indication and use. One other product contains Von Willebrand's factor with factor 8 for that specific deficiency disease. These recombinant protein blood factors replaced blood derivatives and avoid the potential viral contamination and immune reactions that were previously observed in these patients. Three interleukins (ILs) are in use for renal cell carcinoma and malignant melanoma (IL-2), cutaneous T-cell lymphoma (denileukin), and thrombocytopenia associated with cancer chemotherapy (IL-11). These interleukins are protein products that can cause substantial multiorgan toxicity, especially cardiovascular, and limit their full clinical usefulness, which characterizes most interleukins. Denileukin is a fusion protein of IL-2 and diphtheria toxin.

Mabs

The 1990s were called the biological era of Mab proteins, as one product was approved in 1980s that greatly expanded to seven products in 1990s with a wide range of indications. This growth continued into the 21st century: 10 more products in the first five years are now therapeutic products. The nomenclature for Mabs is highly structured and guides the identification of their origin and usage areas. For example, trastuzumab (Herceptin) can be identified as used in cancer and is a "humanized" Mab as follows: tras-tu-zu-mab; 'tras' is the name fragment unique to the Mab protein; "tu" is identified for cancer indications, vs. "li" for inflammatory conditions, and "ci" for a cardiovascular indications or mechanisms of action; "zu" identifies the "humanized" type of Mab protein vs. "mo" for fully murine proteins, "xi" for chimeric proteins (75% human and 25% murine), and "mu"for fully human Mabs; mab of course is for monoclonal antibody. Hence again, Tras-tu-zu-mab is a cancer-humanized Mab.

In addition to treatment of rejection of organ transplants, mab indications include prevention of blood clots, seven cancers, e.g., metastatic breast cancer, non-Hodgkins lymphoma, leukemias, and inflammatory disease of the gastrointestinal system and rheumatology areas, e.g., Crohn's disease and rheumatoid arthritis, neurologic disease, e.g., multiple sclerosis, and viral pneumonia in children. In oncology, Mabs also serve as carriers of radioactive species to enhance cell kill, e.g., tositumomab I-131. The substantial growth in Mab products with major new indications is predicated on the specificity of Mabs to their cell targets and especially on the process of humanization of the murine antibodies, leading to less side effects and more affinity for the target receptors. Mabs are also developed for diagnostic testing with five Mab products in U.S.A. and three in Europe, e.g., OncoScinct for colorectal and ovarian cancer, *carcino-embryonic antigen* (CEA) scan for colorectal cancer, ProstaScinct for prostate cancer, MyoScinct for myocardial infarction, Tecnamab Kl for melanoma, Verluma for lung cancer, LeukoScan for osteomyelitis, and Humaspect for colorectal cancer.

Enzymes

The first protein enzyme developed in the late 1980s was alteplase (t-PA), a thrombolytic agent used to minimize complications owing to blood coagulation in acute myocardial infarction. Six further enzymes were developed for similar indications. The cardiovascular indications related to thrombolysis have expanded to include pulmonary embolism, stroke, percutaneous angioplasty, arterial vessel stenting, and acute coronary syndromes with or without unstable angina. Four of these recombinant protein enzymes were identified from animal sources, as noted in discussions above. A unique enzyme was discovered for cystic fibrosis, dornase alfa, which is the enzyme deficiency responsible for the etiology of this disastrous respiratory disease. This enzyme was the first protein administered by inhalation, given the nature of the disease being a protein deficiency in the lungs. A family of rare single-enzyme deficiencies causes life-ending diseases often in childhood or adolescence, involving multiple organ system disruption. Four such enzyme deficiencies now have the enzyme commercially available for therapy. The enzyme deficiency responsible for *severe compromised immune deficiency* (SCID) is available for replacement therapy too. Finally, a key enzyme is responsible for uric acid metabolism, principal to the eradication of excess nucleic acid material from dying cells especially in cancers; its deficiency leads to hyperuricemia, and is correctable with the enzyme product rasburicase.

Vaccine and Liposome Products

Vaccines in biotechnology employ recombinant techniques to create fractions of a microorganism or virus that retain the activity as an immune stimulant for protection against that infectious species. Hepatitis B vaccines were the first recombinant vaccines created for immunization. Lyme disease now has a preventative vaccine available. A new role for vaccines under study is the use as therapeutic products to treat cancer, wherein the vaccine is specific to a cancer type, turns on the patient's immune system against the cancer, and is used in conjunction with other anticancer treatments. All such products remain in clinical trials. Liposomes are carrier molecules comprising lipids most often in spherical molecules with several layers of lipid, and the drug or biological agent is carried within the lipid molecule. The goal is improved product delivery to target cells, which basically are also lipid sacs, and hopefully less systemic toxicity. Two cancer agents, doxorubicin and daunorubicin, and one antifungal antibiotic, amphotericin, are formulated as liposomes, and their serious toxicities are reduced to some extent, cardiac damage and kidney damage, respectively.

Tissue and Cell Engineering Products

Surgery, especially gastrointestinal and back locations, can lead to complications where abnormal connections can occur between tissues called adhesions. They can be quite persistently painful after surgery and may require a second surgery to eliminate them. Hyaluronic acid products in the form of gels and films are available to prevent them by placement between tissues. The products are biodegradable in situ to non-toxic substances. Also, tissue damage occurs in a variety of diseases where the tissue is accessible for replacement, e.g., skin ulcers from diabetes, pressure, or burns. Skin grafting can be done with exogeneously engineered skin products. Tissue damage is becoming a greater problem as the population ages, and tissues tend to break down more over time in older populations, e.g., osteoarthritis of the knees or facial wrinkling. Knee pain can be relieved and wrinkling reduced with hyaluronic acid products administered directly into tissues, and even chondrocytes can be replaced in the knee. Bone fractures can be mended more rapidly through enhanced processes with biological products or devices that contain bone morphogenic growth proteins (BMP). Facial lipodystrophy in HIV patients can be reduced with a biological product of a form of lactic acid.

Other Biological Products and Categories

A biological carrier (wafer) is used to administer the cancer drug, BCNU, for brain tumors. An antisense molecule (fornivirsen) is used to treat CMV retinitis. The destructive tumor necrosis factor is blocked by etanercept to improve rheumatoid arthritis treatment; its indications were expanded to cover several inflammatory conditions involving varied organs. Another protein, anakinra, attacks the other primary mediator of inflammation, interleukin-1, in arthritides. A fusion protein of a T-cell antigen and the Fc fragment of IgG1, abatacept, was developed to treat rheumatoid arthritis. Psoriasis can be treated now with another fusion protein, alefacept. A peptide of four amino acids is used to treat multiple sclerosis. A peptide treats HIV infections by a unique mechanism of action preventing viral penetration into the cell. Another peptide, ziconatide, from a marine snail is manufactured for severe pain. Two other proteins have been approved for congestive heart failure and sepsis. An enzyme deficiency in emphysema can be treated. An inhibitory protein to growth hormone has been created and pegylated for longer duration of action for acromegaly, pegvisomant. Two antibody fragments are now formulated as antidotes for a snakebite and drug poisoning. These 16 molecules further represent the expanding breadth of indications and new types of molecules for biotechnology products.

Biological Products (Blood Derivatives and Natural Extracts)

Biological products were developed traditionally, before recombinant proteins, as extracts or derivatives of the actual protein from the human body or nature. Albumin is obtained for cardiovascular volume conditions. A fish protein is harvested for osteoporosis. Igs extracted from blood are available for immunodeficiency conditions, viral infections (hepatitis-B and vaccinia), hemolytic anemia in newborns, and idiopathic thrombocytopenic purpurea. Antihemophilic products are still derived from blood. An antiglobulin is produced for kidney transplant rejection. A collagen product and two botulinum toxin products are used for various facial wrinkle problems and cervical dystonia. A bacterial antigen is formulated to enhance immunity to treat a cancer.

2

Manipulating Genes in Field

Plant seeds have evolved to synthesize and accumulate reserve compounds in large quantities. Therefore, transgenic seed crops, particularly the large seeds of cereals and legumes, offer an effective and economical system for the production of biomolecules. Plant systems produce recombinant proteins that are free from animal viruses and other pathogens that could potentially infect humans. This ability to produce safe recombinant proteins opens new perspectives in agriculture using conventional techniques for harvesting and downstream processing. However, public concerns about large-scale, field-grown transgenic plants require the investigation of practical issues such as biosafety and containment. Seeds of fodder pea varieties (*Pisum sativum* L.) appear to be especially suitable as safe bioreactors. Pea plants are self-fertile, with a rate of less than 1% cross-pollination because the flowers remain closed. Pea cultivars are sexually incompatible with most of the wild pea species, and spring cultivars do not survive frost conditions.

Although all plant organs can in principle be used for the expression of foreign genes, seeds appear to be a preferable choice. Therefore, strong and seed-specific gene promoters are essential tools for efficient and controlled transgene expression. The broad bean (*Vicia faba* L.) '*unknown seed protein*' (USP) promoter meets most of the requirements for molecular farming in pea. The molecular basis of seed specificity for this promoter is sufficiently well understood and it has been used to control the expression of various transgenes under greenhouse conditions.

In this chapter we describe the use of pea seeds to express the bacterial enzyme α-amylase. Bacterial exoenzymes like the heat stable α-amylase from *Bacillus licheniformis* are important for starch hydrolysis in the food industry. The enzymatic properties of a-amylase are well understood, it is one of the most thermostable enzymes in nature and it is the most commonly used enzyme in biotechnological processes. Although fermentation in bacteria allows highly efficient enzyme production, plant-based synthesis allows *in situ* enzymatic activity to degrade endogenous reserve starch, as shown in experiments with non-crop plants performed under greenhouse conditions. Finally, the quantitative and sensitive detection of a-amylase activity makes it a suitable reporter gene product to study the strength and specificity of different promoters. This chapter includes results obtained during a two-year field trial, which - to our knowledge - was the first field trial with transgenic pea in Europe. The experiment was performed to test the applicability of a transformation and expression system comprising a vector cassette with a strong, seed-specific promoter, the transformation of a commercial fodder pea variety and stable transgene expression under field conditions during two vegetation periods.

Procedures for Foreign Protein Expression in Transgenic Pea Seeds

Plant Material, Transformation and Field Growth

The experiments described in this chapter involved fodder pea (*Pisum sativum* L.) varieties "Erbi", "Eiffel" and "Power". The transformation method we used was suitable for gene transfer to several different commercial fodder pea varieties and is based on the protocol of Schroeder *et al.* Essential modifications have been described previously. Explants for transformation were cut from the embryonic axis of immature seeds. For explant preparation, the root end of each segment was cut off and the epicotyl and apical meristem regions were sliced transversely into 3–5 segments. The slices were co-cultivated with *Agrobacterium tumefaciens* for 3–4 days at 21°C with a 16-h photoperiod. After co-cultivation, explants were washed 3–4 times with sterile water. After callus induction, developing shoots (20 mm in length), were selected on 10 mg L^{-1} phosphinothricin for several days. The resistant shoots were grafted to a rootstock of "Erbi" seedlings *in vitro*. After 6–10 days, the plants were adapted to soil and grown up in a climatic chamber. When primary transformants reached the 4–6 node stage, leaves were painted with a 0.5% solution of BASTA herbicide. Inheritance of the transgene was also monitored by herbicide application.

Plants from the first six generations after transformation were grown in growth chambers and under greenhouse conditions. The field trial was performed with a homozygous line from generations T_7 and T_8 in spring/early summer 2000 and 2001, on a field of approximately 100 m^2. One hundred transgenic plants and one hundred wild type control plants were grown. To prevent uncontrolled seed distribution by birds, the field was covered with a net. The environmental conditions of the field trial were as follows: the field was 111.5 m above sea level, the average precipitation was 492.1 mm, the average temperature was 8.5°C, and the field comprised highly fertile black earth surrounded by a protective strip planted with *Avena sativa* L. var. *nigra*, *Fagopyrum esculentum* and *Trifolium resupinatum*. The field experiment was performed according to German law with permission from the state of Saxony-Anhalt. To protect plants against pests, compounds such as "Ripcoord", "Karate" and "Perfecthion" were applied during the field trial. Plants were harvested by cutting the stem and whole plants were bagged. Mature seeds were separated from the plants in the laboratory to avoid seed loss during harvesting. To ensure that no transgenic pea plants remained after the field trial, the area was left unused for two further years. During this period, the area was inspected every two months. To date, three plants have been found.

Transformation Vectors and Analysis of Transgenic Plants

The expression cassette contained the 638-bp *V. faba* USP promoter, the *Bacillus licheniformis* α-amylase gene and the *Agrobacterium tumefaciens* octopine synthase (*ocs*) terminator. The cassette was isolated as an *Xba*I fragment and inserted into the binary vector pGPTV-*bar*. Subcloning procedures were carried out according to Sambrook *et al.* A modified protocol was used to determine the α-amylase activity in transgenic pea seeds. All enzyme activities are presented in Ceralpha Units (CU). Pea seeds were milled under liquid nitrogen and 500 mg flour was extracted with 10 ml of extraction buffer B (0.1 M maleic acid, 5.8 g NaCl, 2 mM $CaCl_2 \times 2H_2O$, 0.01% sodium azid, pH 6.5) for 20 min at 42°C. One milliliter of the extract was incubated at 80°C for 30 min and centrifuged at 13000 x g for 10 min in an Eppendorf bench centrifuge. The α-amylase activity was determined in the supernatant. Extracts of seeds younger than 13 days after pollination (DAP) were measured without dilution, whereas seed extracts from later developmental stages were diluted 5 or 10 fold before measurement. Agar plates containing 0.5% starch were used for a qualitative test of α-amylase activity. Ten microliters of extract was spotted on the surface of the plate, incubated overnight at 37°C and the remaining non-hydrolyzed starch was detected with Lugol solution.

Expression of α-Amylase in Transgenic Pea Seeds

Three commercial fodder pea varieties ("Erbi", "Eiffel" and "Power") were transformed using the transgene construct. The transgene copy number was determined by Southern blot hybridization and segregation analysis. The characteristics of each line, in terms of transgene copy numbers and enzyme activities, are summarized in **Table 2.1**. Doubling the copy number by self-crossing and achieving homozygosity resulted in a rather precise doubling of the enzyme activity.

Table 2.1. Number of independent transgenic lines, number of single insertion lines and the range of enzyme activities in pea seeds.

Commercial variety	*Independent transgenic lines*	*Single insertion lines*	*Enzyme activity (CU kg^{-1} seeds)*
Erbi	3	1	4000–9000
Eiffel	4	1	4000–10,000
Power	6	4	4000–6000

The expression profile of the USP promoter was monitored under field conditions using α-amylase activity as a reporter. The earliest sign of α-amylase activity was observed 8 DAP. Although the absolute level remained low, a ~10-fold increase in promoter activity was detected between 12 and 13 DAP. The activity steadily increased during seed development reaching maximum levels of about 6000 CU kg^{-1} in mature seeds.

One homozygous genotype with a single transgene insertion was selected for the field experiment. The enzyme activity in transgenic seeds remained constant over several years without any herbicide

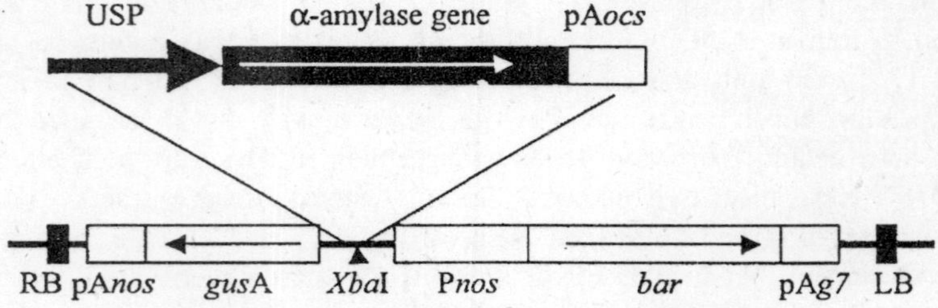

Fig. 2.1. Structure of the expression construct.

selection for the presence of the transgene. No clear difference in α-amylase expression could be detected in plants grown under greenhouse, growth chamber and field conditions. Wild type controls and transgenic lines showed similar yields under all growth conditions. In both 2000 and 2001, the yield (measured as pods and seeds per plant), was found to be approximately 10-fold higher under field conditions. As expected from previous reports, the USP promoter fragment was predominantly seed-specific. The only exception was a low level of transgene expression in pollen, but this was only 1–2% of the level observed in seeds.

Table 2.2. Transgene expression is stable over eight generations and is independent of growth conditions.

Generation	*Year*	*Growth conditions*	*Enzyme activity ($CU\ kg^{-1}$ seeds)*
T_1	1996	Growth chamber	n. d.
T_2	1996/97	Growth chamber	7000
T_3	1997	Growth chamber	6800
T_5	1997/98	Growth chamber	7800
T_6	1998	Growth chamber	7000
T_6	1998	Greenhouse	6500
T_7	1999	Growth chamber	6900
T_7	2000	Greenhouse	6300
T_7	2000	Field trial	7800
T_8	2001	Greenhouse	6500
T_8	2001	Field trial	6200

Table 2.3. Transformed (TG) and non-transformed (WT) plants do not differ in yield. Field conditions result in approximately 5–10 fold higher yields.

	Pods per plant		*Seeds per plant*	
	WT	*TG*	*WT*	*TG*
T_7/2000				
Growth chamber	8.0	7.8	30.4	34.1
Greenhouse	7.6	8.0	22.5	20.4
Field trial	43.4	43.0	220.8	216.0
T_8/2001				
Greenhouse	8.1	6.5	25.3	23.2
Field trial	64.0	61.0	253.0	251.0

The data presented in this chapter demonstrate the potential of pea seeds as bioreactors under field conditions. The chosen transgene was stably inherited and expressed without selection pressure. The expression level remained unchanged over several generations. Apart from presence and expression of the transgene, there was no observable difference between transgenic plants and wild type controls. With the level of transgene expression per seed unchanged, the seed yield per plant was about 5–10 fold higher under field conditions than under greenhouse or growth chamber conditions. One reason for this could be the choice of an established commercial fodder pea variety instead of an experimental laboratory line. No transgenic plants survived over winter, thus providing a built-in biosafety feature.

Based on a seed biomass yield of approximately 4 t $\times$ ha^{-1} for pea in Saxony-Anhalt in the year 1999 and an enzyme activity of 7000 CU kg^{-1} (average of the two field trials), a theoretical yield of 28 $\times$ 10^6 CU ha^{-1} would be expected. This amount clearly exceeds the theoretical enzyme yield calculated for the grain legume *Vicia rarbonensis* (about 10^4 CU ha^{-1}). However, even given the 100-fold improvement in yields achieved when switching from *V. narbonensis* to pea, plants cannot yet match the efficiency of enzyme production in recombinant bacteria. Therefore, enzymes like α-amylase should not be expressed in seeds with the intention of purifying them, since this would not make economic sense. Rather, the potential future advantage of plant-based enzyme production is likely to be the digestion of starch and other substrates *in planta*, as has been shown in *Vicia narbonensis*. This situation is likely to be different for more valuable recombinant proteins, like antibodies, although the yields reported thus far are difficult to compare. Using the seed-specific legumin A promoter, transgenic pea plants produced up to 9 μg per gram fresh weight of a functional single-chain Fv fragment antibody.

There are several approaches that could be used to increase transgene expression levels in seeds. One possibility is the co-expression of a second transgene using another seed-specific gene promoter with a different developmental expression profile. The rather early developmental profile of the USP promoter would be complemented best by promoters whose activity peaked later on in seed development, such as the legumin A promoter of *Pisum sativum* or the legumin B promoter of *Vicia faba*. Alternative promoters have been suggested previously. Another possibility is the use of a longer version of the USP promoter, since extension of the promoter fragment used in this study was found to increase the expression level under greenhouse conditions by a factor of at least two. Finally it would be tempting to use the lines described in this chapter in a conventional breeding program, aiming to a increase the expression level still further.

Based on previous experiments in tobacco and *Arabidopsis*, the USP promoter fragment we used was considered to be highly seed specific. However, thorough examination of the field-grown pea plants revealed some promoter activity in pollen, albeit 100-fold lower than the activity seen in seeds. To test whether or not the pollen expression is restricted to pea, we also examined pollen from transgenic tobacco plants carrying the USP promoter and other seed-specific promoters. Anther activity was observed for the promoters of several seed-specific genes, including those encoding phaseolin, legumin and sucrose binding protein. Anther activity may therefore be a more general property of seed-specific promoters, although the reasons behind this phenomenon are unknown. It may reflect a common regulatory program in somatic cells competent for embryogenesis or a common regulatory program is tissues that undergo desiccation. In any case, the phenomenon needs further investigation, since even the low promoter activity in anthers might lead to a reduction of pollen fertility with transgene products other than the α-amylase.

In summary, we provide an initial technology package for biofarming in pea seeds consisting of a seed-specific promoter cassette for stable transgene expression, an efficient procedure suitable for the transformation of commercial fodder pea varieties and the safe handling of transgenic peas under field conditions. The package was validated by expressing the bacterial exoenzyme α-amylase.

3

Bioelastomerism

The primary function of elastomeric components used by the pharmaceutical industry, which includes both drugs and medical devices, is to protect and deliver. Elastomeric components are typically primary packaging components; that is, they are or may be in direct contact with the dosage form. They must neither interact with the dosage form nor allow the ingress or egress of materials. Elastomeric or rubber components typically provide the means for sealing parenteral containers of varying size and shape because of their unique physical properties. Elasticity particularly permits intimate contact between the closure and the relatively rigid surfaces of container openings. No other material known today has this same unique property. This property is primarily responsible for the ability of the closure to be pierced with a sharp device, such as a hypodermic needle, and then to reseal. This is an example of the delivery function of elastomeric components. Rubber closures are an essential component of the primary package for most parenteral or injectable products.

In order to understand how rubber performs its unique "protect and deliver" function, it is necessary to know how rubber is compounded and manufactured.

Rubber Compounds

Rubber, like all other primary packaging materials, is not inert. Any material used in the compounding of the rubber component may be leached into and/or chemically react with the dosage form. These materials, and the amounts used, are significantly different from those that may be used in industrial rubber belts, tires, and hoses where heat, abrasion, and solvent resistance are typically the main concern.

More than one type of each material may be used in a rubber compound. For example, compounds containing two elastomers, such as isoprene and chlorobutyl rubber, and two pigments, such as titanium dioxide and carbon black, are very common. Compounds used for parenteral closures consist of an elastomer as the base material combined chemically and physically with other necessary rubber chemicals.

Table 3.1. Rubber compounding materials and their function

Material	*Function*
Elastomer	Base material
Curing agent	Forms cross-links
Accelerator	Affects type and rate of cross-links
Activator	Alters efficiency of accelerator
Antioxidant	Antidegradant
Plasticizer	Processing aid
Filler	Affects physical properties
Pigment	Color

The elastomer determines most of the physical and chemical characteristics of a rubber compound. Typical elastomers are natural elastomers such as *natural rubber* (NR), sometimes called crepe, and synthetic elastomers such as butyl (including chlorobutyl and bromobutyl), *ethylene propylene diene monomer* (EPDM), and *styrene butadiene rubber* (SBR).

Table 3.2. Common elastomers used in parenteral packaging components

Elastomer	*% Use*	*Reason*
Butyl/halobutyls	80	Excellent O_2 and moisture barrier
Natural/isoprene	10	Excellent physical properties such as reseal
EPDM	5	Good heat resistance and surface lubricity
Nitrile and neoprene	3	Resistance to vegetable and mineral oils
SBR	1	Blended with isoprene/NR or halobutyls to improve physical properties
Silicone	<1	Excellent heat resistance; poor O_2 and moisture barrier; high cost

In the pharmaceutical industry, natural rubber and its synthetic analog, isoprene rubber, are normally used in products that require good physical strength or that must be able to withstand multiple punctures while maintaining seal integrity. Due the "*latex sensitivity*" issue, the use of natural rubber is declining. Neoprene, a halogenated form of polyisoprene, is typically used for oil-based pharmaceutical products, such as mineral oil or vegetable oil. SBR and nitrile rubber (NBR), which isused primarily for oil-based products, are specialty elastomers that are not as commonly used. Butyls and halobutyls comprise the largest segment of pharmaceutical rubber compounds, having properties that make them applicable for the packaging of many products, especially those requiring protection from moisture vapor or oxygen. Most lyophilized and powdered products require this protection, and some liquid products require protection from oxygen. EPDMs are less commonly used, though they have been selected for large intravenous (IV) stoppers. Silicones also are not often used due to their permeability to moisture vapor and oxygen as well as their relatively high cost.

Curing (or vulcanizing/cross-linking) agents are chemicals used to cross-link elastomer chains into the three-dimensional (3D) network required to give a rubber component the desired elasticity. The term "*vulcanization*" indicates that heat is employed in the manufacturing or molding process. Common curing agents are sulfur, thiurams, zinc oxide, peroxides, resins, and amines. A desirable property of pharmaceutical rubber formulations is "cleanliness," that is, that they contain materials that neither leach nor volatilize into the packaged pharmaceutical. Sulfur-cured rubber, because it requires other chemicals to effect an efficient cure, is not usually as clean as resin-, metal oxide-, or peroxide-cured formulations. The demands of the pharmaceutical industry and regulatory agencies are such that these relatively clean cure systems are becoming more common.

Accelerators reduce the cure time considerably by increasing the cure rate. They are not catalysts because they are chemically altered and, in many cases, also react as curing agents. Common sulfur-cure accelerators are amines, dithiocarbamates, sulfenamides, thiazoles, and thiurams. Some accelerators, because of their reactivity, may form toxic compounds, such as 2-(2-hydroxyethylmercapto)benzothiazole from mercaptobenzothiazole (2-MCBT), residues of which may be extractable. Accelerators that are secondary amines may form toxic nitrosamines.

Activators, which affect the efficiency of accelerators, are commonly added. Normally these are metal oxides, such as zinc oxide or stearic acid. Antioxidants are classified as antidegradants or age resistors. Chemically, antioxidants protect the reactive (sensitive) sites of the rubber chains against oxygen attack. Typical antioxidants are chemicals such as hindered phenols and amines. Unsaturated elastomers, such as natural rubber, require antioxidants for protection against oxidation, which causes surface cracking and loss of elasticity. Saturated elastomers, such as silicones and fluoroelastomers, are resistant to oxidation and usually require no added antioxidants. Some antioxidants are classified as antiozonants, which are designed to provide protection when high levels of reactive ozone are likely to be in the environment.

Plasticizers are used in rubber compounds to assist in the mixing or molding of the rubber, to soften the final vulcanized rubber, or to add surface lubricity to the surface of the rubber component. Examples are paraffinic wax, silicone oil, paraffinic and naphthenic oils, phthalates, and organic phosphates. Silicone oil is commonly used in syringe pistons that must slide freely within a glass or plastic barrel; it also reduces the coring or fragmentation tendency of vial stoppers.

Fillers are materials that modify rubber characteristics (e.g., hardness) and improve its physical characteristics (e.g., tensile strength), in addition to reducing costs. Rubber is sometimes compounded without the use of fillers; the resultant product is called "gum rubber." Typical fillers are calcined and hydrated clays, magnesium silicate (talc), magnesium oxide, and silicas. Carbon black, a common filler used to increase the heat resistance in industrial components such as tires, is not used as a filler in pharmaceutical components but it is used in smaller amounts as a black pigment. *Polynuclear aromatic* (PNA) hydrocarbons are a concern with carbon blacks but the grades used by manufacturers of pharmaceutical components contain very low concentrations.

The pigments used are inorganic salts and oxides, carbon black, or organic dyes that are used for aesthetic or functional purposes (e.g., identification, designating a dosage, etc.). Typical pigments are carbon black, titanium dioxide, and iron oxide. With these three pigments white, black, red, and many shades of gray and pink can be produced. These pigments are chemically pure and stable, non-toxic, and relatively inexpensive. Other pigments such as phthalocyanines and ultramarine blue can be used for blues and greens, but their color fastness in not as good as the aforementioned pigments.

In terms of percentage by weight, the elastomer and filler are the chief materials used, accounting typically for over 90% of a compound. However, the other "minor" materials are quite necessary in order for the compound to have the necessary chemical, physical, and toxicological properties required for a functional packaging component. For example, without the curing agent the compound would remain a physical mixture of materials that would have the consistency of chewing gum. On the other hand, materials such as pigments may be omitted from the compound with only minor consequences—i.e., the loss of the desired color.

Table 3.3. Typical thermoset rubber compound

Material	*Percent by weight*
Chlorobutyl rubber (elastomer)	52.7
Calcined clay (filler)	39.4
Paraffinic oil (plasticizer)	4.4
Titanium dioxide (pigment)	1.1
Carbon black (pigment)	0.13
Thiuram (curing agent/accelerator)	0.14
Zinc oxide (activator)	1.6
2,6-Di-tert-butyl-4-sec-butyl phenol (antioxidant)	0.53

Selection of Compound Materials

Many materials may be used in a rubber compound; however, only a fraction of materials are acceptable in components used for the drug industry. A source of acceptable materials is the U.S. Code of Federal Regulations (CFR). Since the CFR has no list for drug contact, the drug industry uses the CFR list designated for foods. Applicable sections of 21 CFR are as follows:

Section 175—Indirect Food Additives

Sections 177 and 178—Indirect Food Additives Polymers

Sections 182, 184, 185—Generally Regarded as Safe (GRAS) Lists

Section 177.2600—Rubber Articles Intended for Repeated Use

There are some cautions with the CFR list. First, manufacturers may not always submit materials to the Food and Drug Administration (FDA) for listing in the CFR. They may not want to take the time or incur the costs if they see only a limited market for their material in the pharmaceutical or food market. Second, some materials, such as 2-MCBT, which is not permitted by the FDA, are listed but are strongly discouraged. Finally, some materials listed in the CFR, such as food, drug, and cosmetic dyes, are really not applicable for rubber compounds used for pharmaceuticals. Many of these dyes are water-soluble and are not applicable for rubber formulations that come in contact with aqueous solutions, since the dyes could be extracted from the rubber and discolor the drug.

Component manufacturers may use materials not listed in the CFR provided that acceptable toxicity data is available to the reviewing health authority.

Types of Rubber and the Manufacturing Process

Elastomeric closures for parenteral products are made from two types of elastomers or rubbers. Thermoset rubber, the most common, undergoes a chemical reaction during the molding or component-forming processing. In this chemical reaction cross-links, or bonds are inserted between the long polymer chains to form a resilient 3D network. Without these cross-links elastomeric closures would have properties resembling those of chewing gum, which is an uncross-linked rubber blended with sugar, flavors, and food coloring. The cross-linking process is not reversible. Once a closure is molded it cannot be remolded into another shape or size. Addition of heat only causes degradation or reversion of the rubber.

Another type of rubber that is used frequently is thermoplastic rubber. Components are fabricated in a process that is similar to that used for common hard plastics, such as polyethylene or polystyrene, but the final product is an elastic material with properties otherwise equivalent to those of thermoset rubbers. No chemical reactions are involved in the processing of a thermoplastic rubber. The fabrication process consists of heating the rubber compound until it liquefies, injecting the liquid into a mold, cooling the mold, and finally removing the closure from the mold. The process is reversible. Closures can be remelted and remolded into different shapes or sizes as desired. Cross-linking in this case is not a chemical process but a physical intertwining of polymeric chains. The resulting intertwined 3D network gives thermoplastic elastomers their elasticity and resiliency.

Currently, thermoplastics account for less than 5% of the elastomeric closures for parenterals. Their limited resistance to heat deformation under stress during autoclave sterilization is the main reason for this limited use. However, thermoplastics have two advantages over thermosets. First, they are chemically less complex and therefore less prone to interact with parenteral medications, and second, they may be manufactured by a simpler and more automated process. Thermoplastic elastomers have found use in baby bottle nipples and dropper bulbs that are not typically heat sterilized under compression. The manufacturing of rubber components for pharmaceutical applications is a multistep process with controls on each step. Fully validated processes are now common.

There are generally three molding processes used for manufacturing pharmaceutical closures: compression molding, injection molding, and transfer molding. The choice of molding method usually depends on the necessary final dimensional tolerances of the item being molded. Injection molding gives the best dimensional tolerances; however, it is usually the most expensive technique, especially as compared to compression molding. Thermoset rubbers are commonly compression or transfer molded, while thermoplastic rubbers are typically injection molded. Sheets of molded components vary in shape from round to rectangular, and in size from 12 in. in diameter to 36 × 36 inches square, and may contain 50–10,000 components.

A recent trend in rubber component manufacturing is the production of "*preprocessed*" components. Components are prepared for shipment in either one of two states:

1. *Ready to Sterilize* (*RtS*): Typically components are washed, then rinsed with Water For Injection (WFI) to reduce bioburden and endotoxin levels, lubricated with silicone oil, and finally packaged in a classified area (Classes 100–10,000) in Tyvek bags that can be steam sterilized by the drug or device manufacturer before use. Alternately polyethylene bags may be used if the end user is utilizing gamma radiation for sterilization.
2. *Ready to Use* (*RtU*): The process is identical to that used for RtS components except that the component manufacturer takes responsibility for component sterilization and the sterilized components are received by the drug or device manufacturer. No additional component processing is required before use.

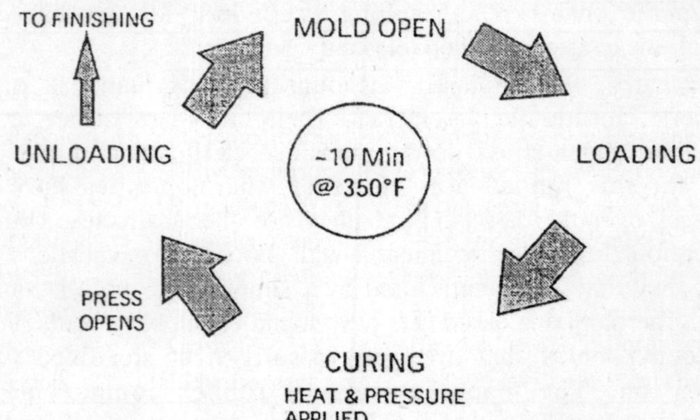

Fig. 3.1. Compression molding cycle.

Both processes typically require extensive validation by the component manufacturer, a description of the process in a Drug Master File (DMF), quality audits by customers, and perhaps FDA approval before RtS or RtU components are utilized. Nevertheless, RtS and RtU are trends that are moving rapidly; most major rubber component manufacturers now market RtS components.

Sterilization of Rubber Components

Heat, radiation, and sterilizing gases may be used to sterilize rubber components. However, components, especially vial stoppers, are most frequently sterilized by pressurized steam (autoclaving), a highly effective method and probably the most reliable of available methods when an F_0 value of at least 8 min is reached. However, the poor thermal conductivity of rubber and the relatively large mass of wet stoppers placed in a stainless steel container for sterilization require a careful validation of the process. Assurance must be obtained that the moist heat penetrates throughout the mass of stoppers so that every stopper receives a dose equivalent to a minimal F_0 of 8 min. However, rubber closures subjected to elevated temperatures or extended heating even at moderate temperatures degrade or revert, which is most frequently evidenced by stickiness.

Therefore, the application of moist heat must be controlled because this degradation is most likely to be greatest at the outer layers of the mass of stoppers. This undesirable effect can be markedly reduced by distributing the stoppers in a shallow tray or otherwise reducing the mass, thus making it possible to reduce the thermal cycle time and the applied heat. Cycle times of 121°C for 30–60 min are usually well tolerated by rubber components made from butyl or halobutyl rubbers. Natural and isoprene rubbers are less tolerant. Increasing the sterilization temperature above 121°C or the time above 60 min is not recommended unless components have been tested to assure that no significant degradation will take place. Dry heat sterilization or drying of components above 105°C is also not recommended without prior testing. Testing may include inspection of the surface for cracks or tackiness, swelling studies to determine state of cure, functional tests such as coring and reseal, and chemical tests such as the United States Pharmacopoeia (USP) protocol or extraction studies performed before and after the sterilization cycle.

Under the usual autoclaving conditions, moist heat does not destroy pyrogens. For closures to be pyrogen-free, a final rinse with WFI before sterilization is necessary.

Ethylene oxide, a sterilizing gas, may be used to sterilize rubber components when they are part of a medical device; however, the gas is readily absorbed by the rubber and sufficient time must be allowed after sterilization for the concentration of residual ethylene oxide to dissipate to acceptable levels.

Radiation sterilization, by either gamma or e-beam, also may be utilized to sterilize rubber components. However, some elastomers, such as butyl, chlorobutyl, andbromobutyl, do not tolerate high doses of radiation, while others, such as natural, isoprene, neoprene, and nitrile rubbers are readily radiation-sterilized without degradation. The challenge with radiation sterilization of rubber components in bulk (i.e., cartons or bags) is to obtain a uniform dose of radiation throughout the entire package—enough of a dose to assure the desired *sterility assurance level* (SAL), usually 10^{-6}, without degrading the components, especially those nearest the radiation source. There are generally two approaches that can be followed. One is to target a uniform dose of 25 kGy, which is generally accepted to assure sterility, while the other is to follow the ANSI/AAMI/ISO 11137: 1994 standard and target a dose of radiation that is dependent on the bioburden of the components. The first approach is usually overkill but may be acceptable for radiation-resistant rubbers. The second approach usually results in lower doses applied, which may be critical with butyl and other rubbers that are not so radiation resistant. There are several necessary steps if the ANSI/AAMI/ISO standard is used. These are:

1. Selection of the SAL, usually 10^{-6}
2. Determination of the average bioburden from samples taken from three consecutive production lots
3. Establishment of the Verification Dose. This is the dose in kGy of the standard that will give a SAL of 10^{-2}
4. Confirmation of the Verification Dose by irradiating 100 components and testing each for sterility. No more than two positives are permitted
5. Establishment of the Sterilizing Dose of the standard
6. Dose mapping, using dosimeters, on a shipping container filled with components. Irradiate and determine the maximum and minimum doses (D_{max} and D_{min}) received in the container
7. Calculation of the Target Dose by multiplying the Sterilizing Dose by the ratio, D_{max}/D_{min}. The Target Dose is typically increased 6% or more to account for dosimeter error and to add additional sterility assurance.

TESTS AND STANDARDS

In-Process Tests

Several tests may be performed on unvulcanized rubber or on standard-shaped test specimens to measure the properties of a rubber compound. These include the following:

1. *Rheometer measurements* measure cure and cure rate characteristics of the rubber. The component manufacturer performs this test on unvulcanized rubber. The rheometer measures the viscosity of the rubber as a function of time at a constant temperature. As time increases, the degree of cure or cross-linking increases and thus the viscosity increases.
2. *Durometer hardness* is measured on tests specimens that meet specific standards for shape and thickness. Durometer hardness is usually measured using the Shore A scale, which measures relative hardness on a scale of 0 to 100 units. Most rubber components for medical use are found in the 35–60 range with 40–50 typical for rubber vial stoppers. Durometer Hardness may be measured on some actual components if they have a sufficiently large flat surface and thickness, i.e., 28–32 mm IV stoppers.
3. *Compression set* is commonly used as a measure of the dimensional recovery of a rubber compound after compression at a defined level, usually 25%, at a specified time and temperature, usually 24 h at 70°C. High compression set values are associated with rubber that "takes a set" or loses its ability to spring back after compression. Low compression set is important for rubber closures and syringe plungers that are heat sterilized while under compression and remain under compression for long periods of time before use but must remain elastic and resilient to maintain seal integrity. Compression set is measured on dimensionally defined test specimens.
4. *Tensile*, *modulus*, and *elongation* are measures of the strength of a rubber compound. They are measured on bow tie-shaped specimens that are clamped in a tensile measuring apparatus and stretched.
5. *Water vapor and oxygen transmission* (WVT and O_2T) are commonly measured on thin-film specimens. Butyl and halobutyl compounds have very low WVT and O_2T rates, while rates for natural and isoprene rubbers are higher and for silicone even higher.

Finished Component Tests

Finished component tests may be divided into three categories: those used as routine identity and/or quality control tests, those tests recommended or mandated by government, standards, and compendial groups, and those test that are part of the larger rubber component acceptance and drug/device approval process. In many instances, a test may fall into more than one category.

Identity and quality control tests

1. *Percentage ash* is a measure of the non-volatile materials in a rubber formulation, such as clays and other fillers.
2. *Specific gravity* is a measure of the type and quantity of fillers in a formulation.
3. An *infrared spectrum* of a compound pyrolysate identifies the elastomer qualitatively (natural, butyl, etc.).
4. An *ultraviolet (UV) spectrum* of an aqueous rubber extract identifies and quantifies the antioxidants, curing agents, accelerators, and other UV-absorbing species extracted from the compound.
5. *Percentage swelling* in an organic solvent is a measure of the degree and consistency of cure or cross-linking that determines the physical and functional properties of a component.

These five tests may be used to characterize a rubber formulation or serve, either individually or in combinations, as quality control tests.

Compendial, standard, and government tests

The most influential of these test protocols are the USP, the *European Pharmacopoeia* (EP), the *Pharmacopoeia of Japan* (JP), the Organization for International Standardization (ISO), and the Parenteral Drug Association (PDA). USP<381>, Elastomeric Closures for Injections, contains five chemical tests and two biological tests. Closures must meet the biological requirements but there are no current specifications for the chemical tests. All USP chemical tests are commonly performed on aqueous extracts but isopropyl alcohol and the drug product vehicle are also permitted. A brief description of the USP<381> tests follows.

Turbidity

The clarity of the closure extract is measured with a nephelometer with appropriate standards. Turbidity is a measure of the insoluble extractables from a closure and is affected by the type and amounts of ingredients in a formulation, pretreatments such as washing, extractions, sterilizations, and degree of cure.

Reducing agents

Organic extractables oxidizable by iodine are determined in this procedure. It is affected by the same variables as turbidity.

Heavy metals

Extractable lead as well as other metals, such as zinc and cadmium, are determined colorimetrically or by atomic absorption.

pH change

The pH of the extract is a measure of the acidic and alkaline water-soluble extractables from a compound.

Total extractables

The sum of the inorganic, organic (non-volatile), soluble, and insoluble extractables is measured in this test. The weight of total extractables is an indication of the "*cleanliness*" of a formulation.

Biological tests

USP 24 lists two levels of biological tests—the in vitro tissue culture test and the in vivo systemic injection and intracutaneous tests. A parenteral closure must pass either type of test to meet USP requirements.

The application of numerical specifications to the USP <381> chemical tests is under discussion and probably will become effective in USP 24 via a supplement. Not all pharmacopeias and standards groups designate the same tests; therefore, it is important to consult a wide spectrum of references to design a test protocol for specific applications and geographical submissions. Of the four protocols compared, the JP is the most stringent in terms of limits for extractables.

The USP, EP, and ISO test protocols are based on a specific closure "area" per volume of water in the extraction step. The JP test protocols, on the other hand, are based on a specific "weight" of closures per volume of water. This difference makes it more difficult for smaller closures than for larger closures in the same rubber compound to meet JP specifications; i.e., 13-mm closures will be less likely to meet JP specifications than 20-mm closures, and 20-mm closures are less likely than 28-mm closures to meet JP specifications.

Although packaging components, such as vial closures, syringe plungers, and needle shields, may meet all compendial and accepted standards, this does not mean that they will be acceptable for use with any specific drug or device. Specific evaluation tests must be performed for that purpose. Compendial and standard tests should be regarded as the first necessary but not sufficient hurdle in the race to gain approval of a component for use with a drug or device.

Specific tests for component evaluation and regulatory approval

There are three requirements for a rubber component:

1. Compatible with the drug
2. Meets functional requirements
3. Provides closure-container seal integrity

Many factors influence the choice of a rubber compound for a particular drug, the most important of which is the solvent vehicle. If the solvent vehicle is an aqueous material, then a butyl, natural, or EPDM may be used. If the solvent vehicle is an oil, then a neoprene or nitrile is utilized.

Configuration is also important and is determined by the required function. A lyophilization stopper, designed to keep out moisture, almost certainly requires a butyl formulation, while a vial stopper for aqueous-based solutions could be formulated from isoprene rubber.

Some preservatives are especially reactive with rubber. Bromobutyl rubbers, but not chlorobutyls, are recommended for drug formulations that contain chlorobutanol. A pH that is either very low or very high affects rubber formulations more than a pH in the range of 5–8. Buffer systems may also affect the choice of rubber formulations. Materials such as phosphates not only attack the rubber at high pHs but also may attack glass as well.

Metallic sensitivities affect compatibility, as drug formulations that are sensitive to divalent cations, such as calcium, zinc, or iron, may not be able to use certain rubber compounds that are cured or pigmented with these materials.

An obvious factor is the need for oxygen or moisture vapor protection. A butyl-based compound is normally chosen whenever protection from materials transmitted either into or from a drug product is required.

When it comes to color preference, most rubber manufacturers of closures used for drugs prefer to use three pigments: iron oxide to produce reds, carbon black for blacks, and titanium dioxide for whites; combinations of these are used for pinks and grays. Organic materials used as pigments are not generally acceptable since they are not very heat stable and are generally more toxic than the three pigments mentioned.

Choice of sterilization method is extremely important in choosing a rubber formulation. Many heat-sensitive drugs, such as proteins, are packaged aseptically; that is, the rubber closure, the vial, and drug are sterilized separately, and then all three items are brought together in a sterile environment to form the final package. The FDA, however, encourages terminal sterilization. In this method, the three materials are brought together and then the entire package, the drug in contact with the vial and closure, is sterilized by heat. This method is much more demanding on the closure than aseptic processing.

Radiation (Co-60) is used to sterilize many rubber items, especially those used in devices. The effect of gamma radiation on rubber closures is a function of the elastomer, dose, and postirradiation time. NR and isoprene are much more resistant to irradiation effects than butyl.

To give the drug manufacturer a high degree of assurance that the closures being investigated for a possible package will be acceptable, a prescreening procedure is commonly used by component suppliers. In this procedure, information and a sample of the drug are obtained from the drug manufacturer. Then three to five possible closures, along with an inert control stopper, are used to package vials of the drug. The closures are put onto the drug-containing vials using an exaggerated closure area-to-volume ratio (2× to 3× the normal ratio). Vials are stored at higher and/or lower temperatures than the drug package will normally experience in the upright, inverted and on-side positions.

An inert control stopper should be used since a drug may be stable against glass but not against uncoated rubber. Only when an inert control is used will it be possible to determine whether the rubber and/or glass vial is the cause of the drug instability. Ampoules do not make good controls for the prescreening of drugs in vials since the glass for vials may be different from that used for ampoules.

The important drug compatibility factors that must be considered when choosing a rubber component for a specific drug/device application. These factors along with the functional and seal integrity requirements can be used to write specifications for components. The packaging engineer, as with the rubber compounder, faces the challenge of choosing a compound that best meets the overall requirements but, only at best, provides a balance of all the actual chemical and functional needs. For example, an engineer may choose an isoprene vial closure because multiple punctures with a large cannula are required. Isoprene will provide the low coring and excellent reseal required. But the engineer may give up some shelf life since isoprene is a poor oxygen barrier. Choosing a butyl closure would give additional shelf life due to its better barrier properties, but coring and reseal would not be acceptable. Closure compound choice is always a give-and-take proposition where drug/device requirements must be prioritized before the choice of a rubber compound can be made.

Table 3.4. Drug compatibility factors affecting the choice of rubber component

Liquid or solid?
Liquid
Aqueous—pH, preservative, buffer cosolvent?
Oil—Vegetable or Mineral?
Solid
Lyophilized or powder fill?
Configuration?
Need for O_2, H_2O, CO_2 protection?
Drug type?
Metallic sensitivities?
Method of closure and package sterilization?
Color preference?

There are four general types of closure–drug interactions:

1. Adsorption occurs when a drug is concentrated at the surface of a closure or vial.
2. Absorption occurs when a drug material is dispersed in the closure matrix.
3. Permeation is the transmission of a drug ingredient through a closure into the atmosphere or transmission of an outside material into the container.
4. Leaching is the process by which closure ingredients are extracted into the drug product.

All four of these interactions commonly occur. No rubber compound is absolutely inert to a drug. In many cases, the extent of the particular interaction is extremely small and may not be measurable, but generally all four are occurring, albeit at a low rate. With proteins, adsorption can be a problem. Many proteinaceous materials made by the biotechnology industry are highly adsorbent onto rubber surfaces and their potency may be readily lost. Other drugs products, especially ones that are very acidic or basic, may attack the stopper and cause the extraction of rubber compound ingredients. Common leachables from rubber closures include low molecular weight elastomer fragments, metal ions, antioxidants, plasticizers, lubricants, curing agents, and accelerators. The rate and relative importance of these four interactions determines the degree of compatibility or incompatibility of a stopper with a drug product.

In May 1999, the FDA issued an updated guidance entitled "Container Closure Systems for Packaging Human Drugs and Biologics—Chemistry, Manufacturing and Controls Documentation," that listed in tables the packaging information that should be submitted in an application. The guidance divides the information into four sections:

1. *Description*. Overall general description of the container-closure system plus specific information on suppliers, materials of construction, and postmanufacturing treatments.
2. *Suitability*. This section prescribes that tests must be done to assure protection of the drug product, safety of the packaging component material, compatibility of the component with the drug product, and performance of the component.
3. *Quality control*. The rubber component manufacturer's release criteria and drug packager's acceptance are found in this section. Also recommended is a method to monitor the consistency in composition of elastomeric components such as periodic extraction profiles.
4. *Stability*. Testing of the drug product using the packaging component is required. Unlike compendial tests that utilize water for extraction studies of the packaging component, the complete drug product is utilized in these stability studies.

This guideline advocates more information on rubber extractables and the proper use of DMFs.

Knowledge of extraction data from elastomeric components refers not only to the broad (and usually non-specific) type of extractable data generated in USP<381> testing, but also to the identification and quantification, where necessary, of specific extractable species. Example of non-specific extractables include turbidity, reducing agents, heavy metals, pH change, and total extractables-; They all measure broad types of extractables. Specific extractables include inorganics, such as zinc or lead and organics, such as PNA, stearic acid, nitrosamines, tetramethylthiuram disulfide, or 2,6-di-*tert*-butyl-4-*sec*-butyl phenol. Both liquid (HPLC) and gas chromatography (GC) are well suited for this purpose. Drug/device producers who utilize elastomeric packaging components require information from their suppliers on extractables (extractable profiles), including test methods, that are generated in water at various pH values (i.e., 3, 7, and 10) and in organic solvents such as isopropanol. Armed with this information, they can look for extractables in their drug products.

Confidential packaging component information, such as the compound recipe, may be placed in a Type III DMF so that the FDA can review the information when it reviews the drug application (IND, NDA, ANDA, or BLA). Most elastomeric compounds are filed at the request of a pharmaceutical manufacturer who has chosen to use the rubber closure in one or more drug packages or devices applications. The name of a rubber compound is associated with a precise recipe that designates specific ingredients and quantities. Once a rubber compound is filed in a DMF, no changes can be made in that compound without changing the DMF and notifying all customers on whose behalf the DMF was accessed and reviewed by the FDA. Since changes in a supplier's DMF may require additional stability studies by the drug manufacturer, changes are infrequent.

Recent Issues and Developments

Latex Sensitivity

There are medical and regulatory issues surrounding the use of "*latex rubber*" due to allergic reactions that have resulted in medical emergencies. Even deaths have been noted. There are two broad types of rubber—natural and synthetic. Several synthetic rubbers or elastomers are used for pharmaceutical components. However, there is only one type of commercial NR, which is derived from the rubber tree *Hevea Brasiliensis*. "*Latex sensitivity*" is associated only with NR and not with the synthetics, although other types of allergic reactions can result from contact with synthetic rubbers. NR is processed and used in two forms—liquid or latex rubber and solid or dry rubber, often referred to as crepe, SMR, and SIR.

Specific proteins that are contained in both the latex and dry types cause the sensitivity to NR. Thus, the medical community and regulatory authorities have used the term "*latex*" for both forms of NR when "*latex sensitivity*" is discussed. Latex reactions reported in the literature have thus far been attributed to contact with components made from liquid rubber but not from dry rubber. Components made from liquid rubber are made by a dipping process that is best suited for thin-walled items such as gloves, condoms, and catheters. None of the typical rubber packaging components is made from latex rubber; —they are all made from dry rubber via a molding process. Many studies have been published regarding the allergic properties of latex and dry rubber and further studies are in progress to determine if exposure to these dry rubber components can cause allergic latex reactions.

There is a great deal of regulatory activity in an attempt to protect the public from unexpected latex reactions. In 1996, the USP proposed a change in section <381> that would prohibit the use of NR in elastomeric closures; however, it rescinded this change in April 1997, stating that closure manufacturers should instead devise latex protein limits. A test for the water-soluble protein content of elastomers, based on an ASTM test, was published as USP <836> but no specifications or limits were proposed. At the same time, the FDA published a final rule that made it mandatory to provide labeling statements on medical devices and packaging components that contain NR. This rule was later amended to exclude combination drug/device and biologic/device products such as rubber stoppers and plungers for prefilled syringes. The uncertainty about the allergic risk from dry NR components and pending regulations have significantly reduced the use of NR in packaging for new drugs and in devices. Substitutes such as isoprene and SBR rubbers are finding increased use.

Preprocessed Components

Preprocessed closures, commonly referred to as RtS or Ready for Sterilization and RtU or Ready for Use, are an unstoppable trend in pharmaceutical packaging. The purpose of preprocessed closures is to reduce total processing costs and improve closure characteristics. Typical RtS closure characteristics are as follows:

Endotoxin Level:	<1 EU/closure
Bioburden Level:	<2 cfu/closure
Silicone Level:	10–40 μg/closure
Visible Particulate:	<20 particles in 25 to 50-μ
Matter:	range; <2 particles in 50 to 100-μ
	range; <1 particle over 100-μ

Coating and Surface Treatments

Although the material science of rubber compounds has greatly improved, drug-closure-interactions and surface lubricity are problematic for packaging engineers. Surface modifications of closures are frequently necessary to meet acceptance standards set by the USP, EP, JP, and ISO as well as the expectations of regulatory authorities. In practice, both liquids and solids are applied to closure surfaces to minimize interactions and improve lubricity. Although these coating and surface treatments may add significantly to the purchase price of closures they often reduce the total drug product cost by providing the following benefits:

1. Increased lubricity, which allows faster processing speeds
2. Decreased drug-closure interactions, which permits the marketing of some products not compatible with uncoated rubber and better quality and longer shelf life for others
3. Decreased particulate matter, which reduces the number of units rejected for visible particulate matter

Container-Closure Seal Integrity

Seal integrity tests can be done by both physical and microbial methods, but historically, sterility testing alone has been used. A 1998 FDA draft guidance discusses the replacement of the sterility test with an appropriate container-closure integrity test in the stability protocol, permitting an alternative to sterility testing for proving the continued capability of containers to maintain sterility. Kirsch et al. published a series of four papers that studied mass spectrometry-based helium leak detection, microbial ingress, and vacuum decay and the correlation between these methods. The PDA also published an updated technical report that provides guidance for evaluating pharmaceutical package integrity, and Guazzo has published an excellent review article that outlines the advantages and limitations of current methods.

Rubber packaging and device components are an important part of the overall medical delivery system. Without innovative packaging systems, modern drugs would not be available today. Advances in drug development have initiated research in new packaging and delivery systems while the availability of innovative packaging has led to the introduction of new drug therapies. Innovation, while containing costs and conforming to regulations, is the challenge for the 21st century.

4

Expressive Molecular Farming

Among the many different agricultural expression systems that can be used for the large-scale production of recombinant proteins, field-grown monocot crops, and more specifically cereals, represent one of the most attractive options. The major advantage of cereals is that recombinant proteins can be targeted to accumulate specifically in the seed, e.g. in the embryo, the aleurone layer or the endosperm. The endosperm is particularly suitable because this tissue has evolved for the accretion of storage products, including proteins. Therefore, it contains a well-developed endomembrane system, within which molecular chaperones and disulfide isomerases help to fold proteins correctly, while the desiccated environment in the mature seed protects the stored proteins from degradation. In the best cases, recombinant proteins expressed in seeds have remained stable and active after storage at room temperature for more than two years. The recombinant protein reaches high concentrations in a small volume of tissue, which facilitates extraction and downstream processing. In terms of processing, other advantages of cereal seeds include the lack of phenolic compounds and the relatively simple proteome. Phenolic compounds, such as the alkaloids present in tobacco and the oxalic acid found in alfalfa, can interfere with processing steps, while the presence of many competing proteins can result in the co-purification of non-target endogenous polypeptides with similar properties to the target recombinant protein. A final advantage of seeds is that proteins restricted to seeds do not interfere with the growth of vegetative organs, so even if they could adversely affect the plant, then normal growth is possible. In terms of biosafety, seed expression limits the exposure of non-target organisms, such as pollinating insects, microbes in the rhizosphere or herbivores feeding on leaves, to biologically active recombinant proteins. Furthermore, three of the cereal crops considered below are self-pollinating, reducing the likelihood of out-crossing to non-transgenic crops and wild relatives.

Four cereal crops have thus far been utilized for the production of recombinant proteins: maize, rice, wheat and barley. It is notable that, despite the attention given to tobacco, oilseed rape and potatoes as major expression systems, the only cultivated crop currently used to produce commercial recombinant proteins is maize. In this chapter, we discuss the relative merits of the four cereal expression systems, consider some of the technical issues surrounding recombinant protein expression in cereal grains, and finally discuss examples of recombinant proteins that have been produced in cereals.

Several factors need to be weighed up when choosing the most appropriate cereal expression host for any given protein. These include geographical considerations for crop growth and harvest, the ease of transformation and regeneration, the annual yield of seed per hectare, the yield of recombinant protein per kilogram of seed, the producer price of the crop, the percentage of the seed that is made up of protein and, inevitably, intellectual property issues. Together, these determine the overall cost of production.

Cereal Production Crops

All four of the cereal production crops have advantages for molecular farming, but maize was the first to be developed into a platform expression system thanks largely to the efforts of scientists. Maize was chosen over the other cereals because it has the largest annual grain yield (approximately 8300 kg ha^{-1}) but also a relatively high seed protein content (10%), and these properties together offer the highest potential recombinant protein yields per hectare. Maize is also relatively easy to transform and manipulate in the laboratory, while its short generation interval (normally about 17 weeks) facilitates rapid production scale-up in the field. Maize is the most widely cultivated crop in North America, so the complex infrastructure for growing, harvesting, processing, storing and transporting large volumes of corn seed is already in place. Prodigene currently has two commercial products that are expressed in maize and is developing further lines for the production of a range of pharmaceutical and technical proteins, including recombinant antibodies, vaccine candidates and enzymes.

Rice has many advantages in common with maize including the high grain yields, the ease of transformation and manipulation in the laboratory, and the capacity for rapid scale up. At 6600 kg ha^{-1}, the annual grain yield of rice is slightly lower than that of maize, and the seed also has a slightly lower protein content (8%). Importantly, however, rice has emerged as the model cereal species, and is one of the few terrestrial plants to benefit from a completed genome sequence. Therefore, many useful expression cassettes have been developed based on endogenous rice genes, including constitutive actin promoters, glutelin promoters for seed-specific expression and the α-amylase inducible promoter system which is often employed in rice suspension cell cultures. The major disadvantage of rice compared to maize is that the producer price is significantly higher, so it may be excluded as a mainstream expression host in the West simply on economic grounds. However, the story may be different in Asia and Africa, where rice is traditionally grown. Rice, along with barley has been adopted as a platform production technology by Ventria Bioscience.

Wheat has been used only rarely for molecular farming, probably because the technology for gene transfer and regeneration is not so well advanced as in other cereals. The major advantage of wheat for the commercial production of recombinant proteins is its very low producer price compared to maize and rice. The wheat grain also has a higher protein content than these other cereals (>12%). Unfortunately, wheat also has a much lower grain yield than maize and rice (2800 kg ha^{-1} yr^{-1}) which means that more land is required to produce the same amount of protein. While some recombinant proteins, including pea legumin, have been produced at high levels in wheat endosperm, only low yields have been achieved with antibodies, although this situation may change in the future if better expression cassettes can be developed.

Barley has much in common with wheat as a host for molecular farming. It has a low producer price, a high seed protein content (about 13%) and the technology for gene transfer and regeneration is not widely disseminated. Unlike wheat, however, there have been some very encouraging reports of high recombinant protein yields in transgenic barley, including a diagnostic antibody that accumulated to 150 μg g^{-1} seed weight and a recombinant cellulase that accumulated to 1.5% total seed protein. Further recombinant proteins will need to be expressed in barley before its performance as an expression platform can be judged against maize and rice.

TECHNICAL ASPECTS OF MOLECULAR FARMING IN CEREALS

Cereal Transformation

Until the 1990s, cereals were thought to be outside the host range of *Agrobacterium tumefaciens*, so researchers concentrated on alternative transformation methods. Initially, attempts were made to regenerate cereal plants from transformed protoplasts following PEG-mediated DNA transfer or electroporation. After successful experiments using model dicots, protoplast transformation was achieved in wheat and the Italian ryegrass *Lolium multiflorum*. In each case, transgenic callus obtained but it was not possible to recover transgenic plants, probably because monocot protoplasts loose their competence to respond to tissue culture conditions as the cells differentiate. In cereals and grasses, this has been addressed to a certain extent by using embryogenic suspension cultures as a source of protoplasts. Additionally, since many monocot species are naturally tolerant towards kanamycin, the *npt*II marker used in the initial experiments was replaced with alternative markers conferring resistance to hygromycin or phosphinothricin. With these modifications, it has been possible to regenerate transgenic rice and maize of certain varieties with reasonable efficiency. However, the extended tissue culture step is unfavorable, often resulting in sterility and other phenotypic abnormalities in the regenerated plants.

An alternative procedure for plant transformation was introduced in 1987, involving the use of a modified shotgun to accelerate small (1–4 μm) metal particles into plant cells at a velocity sufficient to penetrate the cell wall ($\sim$250 m s^{-1}). The original device was gunpowder-driven and rather inaccurate. This has been replaced by a commercially-available pressurized helium gun, resulting in greater control over particle velocity and hence greater reproducibility of transformation conditions. In addition, an apparatus based on electric discharge has been useful for the development of variety-independent gene transfer methods for the more recalcitrant cereals. There appears to be no intrinsic limitation to the scope of particle bombardment since DNA-delivery is governed entirely by physical parameters. Many different types of plant material have been used as transformation targets, including callus pieces, cell suspension cultures and organized tissues such as immature embryos and meristems. Almost all of the commercially-important cereals have been transformed by particle bombardment, including maize, rice, wheat and barley.

During the 1980s, it was shown that asparagus and yam could be infected with *A. tumefaciens* and that tumors could be induced. In the latter case, an important factor in the success of the experiment was pre-treatment of the *A. tumefaciens* suspension with wound exudate from potato tubers (*A. tumefaciens* infection of monocots is inefficient because wounded monocot tissues do not produce phenolic compounds such as acetosyringone at sufficient levels to induce *vir* gene expression). Attention then turned towards the cereals. The first species to be transformed was rice. The *npt*II gene was used as a selectable marker, and successful transformation was demonstrated both by the resistance of transgenic callus to kanamycin or G418, and the presence of T-DNA in the genome. However, these antibiotics interfere with the regeneration of rice plants, so only four transgenic plants were produced. The use of an alternative marker conferring resistance to hygromycin allowed the regeneration of large numbers of transgenic japonica rice plants, and the same selection strategy has been used to produce transgenic rice plants representing the remaining subspecies, indica and javanica. More recently, efficient *Agrobacterium*-mediated transformation has become possible for other cereals discussed in this chapter: maize, wheat and barley.

The breakthrough in cereal transformation using *A. tumefaciens* reflected the recognition of several key factors required for efficient infection and gene transfer. The use of explants containing a high proportion of actively dividing cells, such as embryos or apical meristems, was found to increase transformation efficiency greatly, probably because DNA synthesis and cell division favor the integration of exogenous DNA. In dicots, cell division is induced by wounding, whereas wound sites in mono- cots tend to become lignified. Hiei *et al.* showed that the co-cultivation of *A. tumefaciens* and rice embryos in the presence of 100 mM acetosyringone was a critical factor for successful transformation. Transformation efficiency can be increased further by the use of vectors with enhanced virulence functions. The modification of *A. tumefaciens* to boost virulence has been achieved by increasing the expression of *vir*G (which in turn upregulates the expression of the other *vir* genes) and/or the expression of *vir*E1, which is a major limiting factor in T-DNA transfer. Komari *et al.* used a different strategy, in which a portion of the virulence region from the Ti-plasmid of supervirulent strain A281 was transferred to the T-DNA-carrying plasmid to generate a so-called superbinary vector. The advantage of the latter technique is that the superbinary vector can be used in any *Agrobacterium* strain.

Expression Construct Design

The most important aspect of construct design for molecular farming in cereals is the promoter used to drive transgene expression. In dicots, the cauliflower mosaic virus (CaMV) 35S promoter is the most popular choice because it is strong and constitutive, and therefore drives high-level transgene expression in any relevant organ, including seeds. The promoter can be made even more active by various modifications, such as duplicating the enhancer region. Even so, this promoter shows a very low activity in cereals.

Although some plant promoters appear to be just as active in both dicots and monocots, most promoters that are active in dicots need to be modified in some way before they work efficiently in cereals. One general type of modification that appears to work well with a range of promoters is the addition of an intron, usually in the untranslated region between the promoter and the initiation codon of the transgene open reading frame. Several different introns have been used to modify the CaMV 35S promoter and this has led to improved promoter activity in all four of the cereal host species discussed above, as well as in some grasses. For example, four monocot introns were tested with the CaMV 35S promoter in transgenic maize and bluegrass by Vain et al., specifically those from the *Adh1*, *Sh1*, *Ubi1* and *Act1* genes, as well as a dicot intron (*chsA*). In this comparison, the *Ubi1* intron provided the highest level of enhancement in both species on average, but the dicot *chsA* intron perhaps surprisingly achieved a nearly 100-fold enhancement. Other monocot and dicot introns that have been tested with the CaMV 35S promoter include those from the *Bz1*, *cat1* and *Wx* genes. The activity of another dicot promoter, the potato *pin2* promoter, was greatly enhanced in transgenic rice by inserting the first intron from the rice *Act1* gene. More recently, Waterhouse and colleagues adapted their novel pPLEX series of expression constructs, which are based on regulatory elements from *subterranean clover stunt virus* (SCSV), to be used in monocots. This was achieved by adding either the *Ubi1* or *Act1* introns, as well as GC-rich enhancer sequences from *banana bunchy top virus* (BBTV) or *maize streak virus* (MSV). An alternative to dicot promoters enhanced with introns is to use constitutive monocot promoters, of which the maize *Ubi1* promoter (with first intron) is the most widely utilized.

Various seed-specific promoters, many derived from seed storage protein genes, have been employed to restrict recombinant protein expression to different parts of the seed. Early examples include the maize *zmZ27* zein promoter, a maize *Waxy* promoter and the rice small subunit ADP-glucose pyrophosphorylase promoter, all of which are endosperm-specific in maize, and the rice glutelin 1 (*Gt1*) promoter, which is endosperm-specific in rice and maize. More recently, recombinant proteins have been expressed in maize using the embryo-preferred globulin-1 promoter, in rice using an endosperm-specific globulin promoter and two aleurone-specific promoters from barley, and in barley using the wheat *Bx17* high-molecular-weight glutenin promoter. Care must be taken, however, because although these promoters are often described as 'seed-specific', a low level of activity may be present in other tissues. One pertinent example is the maize *Waxy* promoter, which is endosperm-preferred but shows a low level of activity in pollen.

In addition to an active promoter, a strong polyadenylation signal is required for transcript stability. For molecular farming in cereals, terminators derived from the CaMV 35S transcript, the *A. tumefaciens nos* gene, and the potato *Pin*II (protease inhibitor II) gene have been popular choices. The structure of the 5' and 3' untranslated regions should be inspected for AU-rich sequences, and these should be removed where possible. Such sequences can act as cryptic splice sites, cryptic polyadenylation sites and mRNA instability elements. Some sequences, such as the 5' leader of the petunia chalcone synthase gene and the 5' leader of tobacco mosaic virus RNA (also known as the omega sequence) have been identified as translational enhancers although they may not always work effectively in cereals. Further important factors that influence translation include the presence of a single AUG codon within a consensus Kozak sequence, since multiple AUG codons, where present, often result in pausing and inefficient translational initiation. Different species also have very different codon preferences when specifying degenerate amino acids. Taking the amino acid arginine as an example, the codon CGU is preferred in alfalfa, and is 50 times more likely to occur than the rarest codon, CGG. In contrast, both of these codons are equally prevalent in maize, but the preferred choice is CGC. The expression of foreign transgenes in cerelas can therefore be very inefficient if infrequently-used codons predominate.

One critical consideration for the improvement of protein yields is subcellular protein targeting, because the compartment in which a recombinant protein accumulates strongly influences the interrelated processes of folding, assembly and post-translational modification. For protein accumulation in cereal endosperm, the most important destinations are the protein storage organelles, i.e. the protein bodies and protein storage vacuoles, since these have developed to facilitate stable protein accumulation. In rice, it has been shown that the two major classes of storage proteins, prolamins and glutelins, accumulate in different storage compartments, and are sorted in distinct ways. Prolamins accumulate in protein bodies inside the rough *endoplasmic reticulum* (ER), and these eventually bud off to form separate organelles, whereas glutelins accumulate in protein storage vacuoles, which are derived from the smooth ER, and are conveyed to these organelles by transport vesicles budding from the Golgi apparatus. There have been few studies of recombinant protein localization within the endosperm, but in one interesting report a recombinant antibody targeted to the secretory pathway in rice endosperm, and tagged for retrieval to the ER with a KDEL tetrapeptide tag, was found mainly in protein bodies and to some minor extent also in protein storage vacuoles. The lack of a KDEL sequence in endosperm-expressed proteins generally results in secretion to the apoplast, where in most cases the recombinant protein accumulates in the space under the cell wall. A variety of signal peptides has been used to target proteins to the secretory pathway, including endogenous signal peptides from the transgene (e.g. immunoglobulin signal peptides for recombinant antibodies) and heterologous plant signal peptides such as the barley α-amylase signal sequence (BAASS).

Production Considerations for Cereals

Most of the useful information concerning the practical and commercial aspects of molecular farming in cereals has come from ProdiGene Inc., which has conducted detailed studies into the economic aspects of molecular farming in maize. The development of a product line takes about 30–35 weeks, including 4–6 weeks for vector construction, 5–7 weeks to identify transformants, 4–6 weeks for the regeneration of plants and transfer to greenhouse, and 12–17 weeks for the T_0 plants to reach maturity and set seed. At this point, T_1 seeds are removed for molecular analysis of the integrated transgene, in order to select transgenic events for further characterization. After a further 17 weeks, the T_1 plants are mature and seeds can be removed for protein extraction and scale-up. At this point, about 10 mg of protein can be obtained for preliminary analysis. After another 17 weeks, the T_2 plants are mature and T_3 seeds can be removed for analysis and scale-up. It is routine to obtain about 1 g of pure recombinant protein by this time, but theoretical yields can be up to 1 kg. Also at this time, molecular, genetic and biochemical analysis of samples makes it possible to establish a master seed line for future production. After 12–18 months, the T_3 generation is mature, and it should be possible to obtain 100 g of recombinant protein on a routine basis. Production lines can then be optimized and established in the field.

Another practical issue that needs to be addressed is the compliance of cereal- based production with regulatory guidelines established by bodies such as APHIS and the USDA. Again, ProdiGene Inc. has taken a leading role, developing a procedure in which the producer crop is isolated from other crops, and identity preservation is used from planting through to product extraction, to prevent mingling and contamination. For pharmaceutical products, further strict regulations govern quality assurance and quality control.

Examples of Recombinant Proteins Produced in Cereals

Molecular farming is often subdivided into pharmaceutical and industrial components, the former dealing with medically relevant proteins (e.g. human blood products, antibodies, vaccine candidates) and the later with bulk enzymes, technical proteins and biopolymers. However, there is no strict boundary between these two areas, while other products do not fit clearly into either category. What is clear is that a large and increasingly diverse spectrum of recombinant proteins is being produced in cereals, mostly as laboratory or greenhouse experiments, but also in field trials and large-scale commercial enterprises.

ProdiGene and Maize

The first proteins from transgenic plants to reach commercial status were avidin and β-glucuronidase (GUS) both of which are used as diagnostic agents in molecular biology. An important principle demonstrated by these case studies is that molecular farming in cereals can be an economical alternative even when the natural source of a protein is abundant (i.e. egg whites for avidin, and *Escherichia coli* for GUS) and where a market is already established.

Avidin is an abundant glycoprotein that is routinely purified from hens' eggs. It is widely used in molecular biology due to its very high affinity for biotin. Recombinant avidin was expressed in transgenic maize to establish whether the molecular farming approach could compete with the established source. In the best-expressing transgenic line, avidin represented over 2% of the aqueous protein extracted from dry maize seeds, equivalent to approximately 230 mg per kg of transgenic seed. The *Ubi1* promoter was used to drive transgene expression and the protein was targeted to the secretory pathway using the BAASS. Consequently, the mature protein accumulated in the intercellular spaces. The recombinant protein was shown to be nearly identical to native avidin in terms of molecular weight and biotin binding properties. The protein remained stable in seeds stored at 10°C and was not affected by commercial processing practices (dry milling, fractionation and hexane extraction). Interestingly, avidin expression in transgenic maize plants correlated with partial or complete male sterility. The cost of producing avidin by molecular farming in maize is estimated to be only 10% of the cost of extraction from egg whites.

The β-glucuronidase enzyme (GUS) is widely used as a screenable marker in transformation and promoter analysis, so unlike avidin, its expression and function has been confirmed in virtually all plant species that have been transformed. The use of plants as a commercial source of the enzyme was first demonstrated by Witcher et al.. The *Ubi1* promoter was used to drive transgene expression and the protein, lacking a targeting signal, accumulated in the cytosol. In the best maize lines, the recombinant enzyme accounted for up to 0.7% of water-soluble protein extracted from dry seed, equivalent to about 80 mg per kg dry seeds. Purified recombinant GUS was similar in molecular mass, physical and kinetic properties to native GUS isolated from *E. coli*. Transgenic seed containing recombinant GUS could be stored at an ambient temperature for up to two weeks and for at least three months at 10°C without significant loss of enzyme activity, and could be subject to normal maize processing practices as discussed above without loss of GUS activity.

Transgenic maize lines have also been used to investigate the large-scale production of a variety of other products, although none has yet reached commercial status. Perhaps the closest to commercial production are the protease trypsin, the protease-inhibitor aprotinin and fungal laccase (which is used to modify lignin). The aprotinin gene, driven by the *Ubi1* promoter, was introduced into maize embryos by particle bombardment. The protein was targeted to the extracellular matrix, and in the best-expressing lines, seeds accumulated recombinant aprotinin at levels up to 0.7% soluble protein. Biochemical analysis of the purified recombinant protein revealed similar properties in terms of molecular weight, N-terminal amino acid sequence, isoelectric point, and trypsin inhibition activity as native aprotinin. The cost of aprotinin production in transgenic plants is comparable to the cost of extracting the protein from its natural source. Trypsin and laccase have been expressed in maize using constitutive and seed-specific promoters. Trypsin expression presents a difficulty because the enzyme is a protease, so it could degrade endogenous proteins and itself (autoproteolysis) thus reducing yields. Therefore, the enzyme was expressed as an inactive precursor, trypsinogen, using the seed-specific promoter, a strategy which is patented by ProdiGene. The laccase 1 isozyme from *Trametes versi color* reached the highest expression levels (about 1% total soluble protein) when driven by the embryo-preferred globulin promoter, combined with the BAASS to target the protein for secretion to the cell wall matrix.

The production of a secretory antibody in maize has also been reported. Immature maize embryos were transformed with *A. tumefaciens* containing five trans- genes, encoding the four components of the antibody (heavy chain, light chain, joining chain, secretory component) and a selectable marker. The four antibody trans- genes were driven by the *Ubi1* promoter in combination with the BAASS, and the transgenic seeds were analyzed by enzyme-linked immunosorbent assay to detect assembled antibodies, showing that the antibody accumulated to up to 0.3% total soluble protein in T_1 seeds. Based on ProdiGene's success with other products, the selection of high-performance lines and backcrossing should allow this yield to be increased as much as 70-fold over six generations. Maize has also been investigated as a potential source of oral recombinant vaccines, by expressing the Lt-B protein of enterotoxigenic *E. coli*.

Recombinant Proteins Expressed in Rice

Rice grains were used only occasionally for the expression of recombinant proteins until this system was adopted as a production platform by Ventria Bioscience. In early studies, rice plants were used for the expression of α-interferon and recombinant antibodies, while more recently a cedar pollen allergen has been expressed in grains. The PI' promoter was used to drive α-interferon expression, the antibodies were expressed using either the maize *Ubi1* or enhanced CaMV 35S promoters (with the maize promoter showing five times the activity of the viral promoter), while the cedar pollen antigen was expressed using the seed-specific *GluB-1* promoter, leader and signal peptide. This construct was expected to direct the recombinant protein to type II protein bodies, but instead it accumulated in type I protein bodies, suggesting some competition between the signal sequence and the structure of the recombinant protein in determining its final destination.

In 2002, Ventria Bioscience published two papers describing the expression of human proteins in rice grains. Lysozyme was expressed under the control of the glutelin 1 promoter, leader and signal peptide, and an expression level of 0.6% dry weight (equivalent to 45% of soluble proteins) was achieved. As was the case for maize production lines, the superior transgenic events were selected and maintained over several generations, and showed no loss of expression. Furthermore, the product itself was similar to the native enzyme under all tests and was shown to be active against a laboratory strain of *E. coli*. Lactoferrin was expressed using the same regulatory elements as above and accumulated to a level of 0.5% dry weight of dehusked grains. As for lysozyme, all biochemcial and functional tests showed that the protein was similar in its properties to the native enzyme, and receptor-binding activity in human Caco-2 cell cultures was conserved.

In a further interesting development, Yang *et al.* describe the expression, in rice grains, of a REB transcriotion factor (rice endosperm bZIP), which acts upon the globulin (*Glb*) promoter. When the *Glb* promoter was placed upstream of the *gus*A gene for β-glucuronidase, GUS activity was 2–2.5 times higher in the presence of REB in a cell-based transient assay. This enhancement was abolished when the upstream promoter motif GCCACGT(A/C)AG was deleted from the *Glb* promoter, and a gain of activity was observed when this sequence was added to the normally unresponsive *Gt1* promoter. In transgenic rice grains expressing lysozyme under the control of the *Glb* promoter, the presence ofREB resulted in a 3.7-fold increase in lysozyme accumulation.

Recombinant Proteins Produced in Wheat

As discussed above, wheat has been used only rarely for molecular farming. Thus far, the only example of a pharmaceutical protein produced in wheat is a single chain Fv antibody, which was expressed using the *Ubi1* promoter and achieved a maximum expression level of 1.5 $\mu g\ g^{-1}$ dry weight. Transgenic wheat producing *Aspergillus* phytase has also been reported.

Recombinant Proteins Produced in Barley

There are few reports of barley used for the production of recombinant proteins but some of those reports show high expression levels which, due to the low producer price of barley, could make this a viable production crop. In early reports, recombinant glucanase and xylanase were expressed at very low levels (0.004%) but as stated above, a recombinant diagnostic antibody has been expressed at levels exceeding 150 $\mu g\ g^{-1}$, and a recombinant cellulase enzyme was expressed at levels exceeding 1.5% total seed protein.

Cereal expression systems are among the most advantageous for field-based recombinant protein production, since they combine intrinsic biosafety features (self-pollination in rice, barley and wheat, seed-specific protein expression) with practical benefits, particularly the tendency for recombinant proteins expressed in seeds to remain stable and active for prolonged periods at ambient temperatures. The commercial success of the first maize-derived recombinant proteins produced by ProdiGene has demonstrated that plants, and cereals in particular, represent an economically viable production system which provides a real alternative to mammalian cells, microbial cultures and indeed other crop systems such as tobacco, oilseed rape and potato. It is likely that rice and barley will be the next crops to emerge as commercial production platforms, particularly through the efforts of Ventria Bioscience.

5

BIOVALIDATION

The science that underpins steam sterilization is well known and has been long established. It is the preferred method of sterilization in the pharmaceutical industry; it is used for sterilization of aqueous products in a wide variety of presentations, for sterilization of equipment and porous materials required in aseptic manufacture, in microbiology laboratories for sterilizing media and other materials, and for sterilization of "*massive*" systems of vessels and pipework [steam-in-place (SIP) systems]. Numerous rules and guidelines have been published on the topic, yet steam sterilization and particularly bio-validation of steam sterilization is still a subject for controversy and debate.

The purpose of this article is to reexamine the bio-validation of steam sterilization, to clarify what is needed and why it is needed, and to distinguish the scientific need from the regulatory need in areas where they may appear to differ.

PRINCIPLES

Micro-organisms are inactivated when metabolically irreversible deleterious intracellular reactions occur. At high temperatures and in the presence of moisture, as in steam sterilization, the energy input from the steam inactivates microorganisms by denaturation of intracellular proteins.

Although these reactions are complex at a biochemical level, their kinetics approximate to reactions of the first order. Thus, the kinetics of inactivation of populations of pure cultures of microorganisms take the typical exponential form of reactions of the first order. What this means in experimental practice is that there is a linear relationship when numbers of microorganisms held at high temperatures are plotted on a logarithmic scale against time plotted on an arithmetic scale. There are two highly significant points to be drawn from the kinetics of inactivation of microorganisms.

First, logarithmic scales never reach zero. This means that there can never be any specifications for temperature and time which can guarantee that all microorganisms contaminating items are going to be inactivated. However, the consequences to patients of micro-organisms surviving in allegedly sterile pharmaceutical preparations can easily be fatal. Thus, sterilization processes must be specified to ensure that the probability of microorganisms surviving in treated items is low enough to ensure patient safety. The accepted low probability indicated in the pharmacopeias is that there should be not more than one chance in one million of viable microorganisms surviving on a treated item. This is called a probability of non-sterility of 10^{-6} or a sterility assurance level (SAL) of 10^{-6}.

Second, the inactivation curve takes a regular form. This means that steam sterilization is a predictable process as long as some information is available (or can be safely assumed) about the numbers and thermal resistances of the microorganisms contaminating items before treatment. This is important because there is no practical way to test for the achievement of SALs of 10^{-6}. The sterility or non-sterility of items cannot sensibly be confirmed in a treated item except by sacrificing the item. The pharmacopeial *test for sterility* is a sacrificial test with statistical limitations which have been so extensively criticized over so many decades that they should now be well

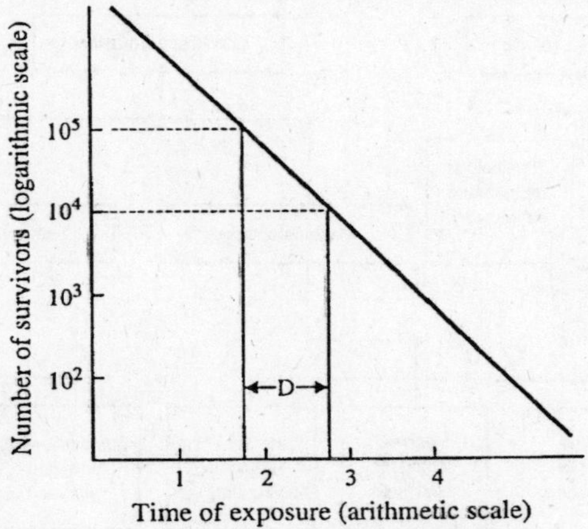

Fig. 5.1. Exponential inactivation of microorganism.

understood. For instance, the sample of 20 items which is generally required in the test would allow a batch containing non-sterile items at a frequency of 1:100 to be passed on four out of every five occasions. This falls a long way short of being able to detect deviations from a standard of not more than one non-sterile item in one million.

Justification of the reliable achievement of SALs of 10^{-6} for particular pharmaceutical items treated according to particular specifications of temperature and time in particular sterilizers is predicated on the regularity and predictability of steam sterilization processes. The means of justification are through scientifically based development of sterilization specifications and sterilizer parameters, and through subsequent validation of the specified processes.

Development of Sterilization Specifications and Sterilizer Parameters

The development of sterilization specifications differs from the development of sterilizer parameters. Both differ from validation.

Pharmaceutical Products and Materials for Aseptic Manufacture—Sterilization Specifications

For pharmaceutical products and materials used in connection with aseptic manufacture, sterilization specifications apply to conditions of temperature and time, or F_0, or combinations of F_0, temperature and time to which the contaminating microorganisms themselves must be exposed over the "hold" period of the sterilization process. In practice, this means actually within aqueous products, on the surfaces of rubber

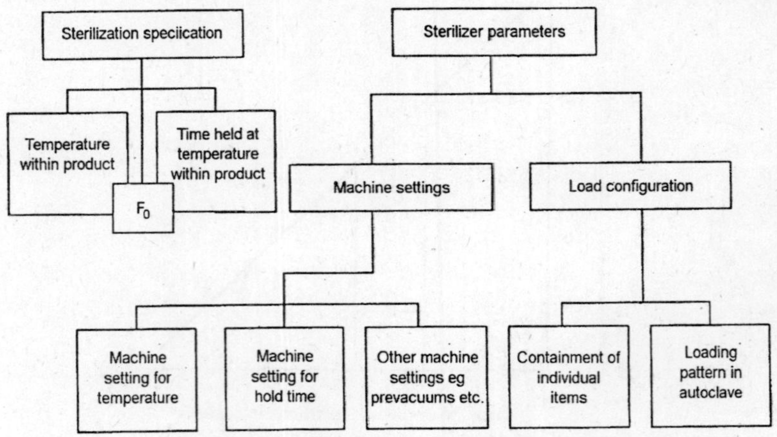

Fig. 5.2. Sterilization specifications and sterilizer parameters.

stoppers or metal machine parts, or within the folds of cartridge filters, etc. The sterilizer parameters are the practical criteria that must be specified to ensure that the sterilization specification is delivered to all parts of the load. They always include specifications for temperature and time, but it is important to recognize the distinction between sterilizer parameters applying to the machine settings on the autoclave console, and sterilization specifications applying to actual conditions within the load. Essential sterilizer parameters also include other specifications, e.g., for load configuration, number and depth of prevacuums, cooling characteristics, etc.

Sterilization specifications are product specific. Sterilizer parameters are specific to combinations of product, presentation, and autoclave. Sterilization specifications may be determined from theoretical considerations or from laboratory data and arewithin reason transferable from presentation to presentation, e.g., from 1 ml ampules to 5 ml vials to 50 ml bags. The sterilizer parameters required to deliver the sterilization specification to these presentations differ within the same autoclave and from one autoclave to another according to differences in load configurations, chamber size, steam entry points, control systems, etc. Sterilizer parameters are not transferable and must be developed empirically for each autoclave.

Sterilization specifications should be easy to develop. The pharmacopeias allow sterilization specifications to be developed from a basis of no actual data concerning the numbers and thermal resistances of microorganisms actually contaminating the items to be sterilized. Although this statement may appear initially to be barren of scientific reason, this is in fact not the case. What the pharmacopeias provide are either a recommended overkill specification (*PhEur*) or principles for specification development (USP) that incorporate amounts of thermal lethality well in excess of that which could ever be practically required to obtain SALs of 10^{-6}—for this reason they are called "overkill" specifications.

In the *European Pharmacopoeia* (PhEur), a specification of 121°C for 15 min is given as the reference condition for overkill sterilization of aqueous preparations. The United States Pharmacopeia (USP) defines a lethality input of 12D.

It is worth considering the thermal resistances of micro-organisms found in pharmaceutical manufacturing environments. It is extremely rare for anyone to have isolated thermally resistant bacteria with D_{121}-values (in water) of greater than 0.3 min. The author of this paper has experience of having determined a D_{121}-value (in water) of 0.8 min for an environmental isolate of *Bacillus coagulans*, but this was several decades ago and was done with what would now be considered fairly primitive equipment. It is therefore probably quite reasonable to assume a worst case D_{121}-value of 1 min. Given this "*worst case*," the PhEur overkill specification of 121°C would deliver 15 decimal reductions which are equivalent to assuring a 10^{-6} SAL for contaminating populations per item of up to 10^9 microorganisms each with a D_{121}-value of 1 min. The USP specification of 12D under the same assumption ensures an SAL of 10^{-6} for populations of up to 10^6 micro-organisms per item.

Thus, the pharmacopeial overkill specifications provide considerable degrees of assurance that SALs of 10^{-6} will be achieved. However, these high theoretical levels of overkill are contingent upon D-values in water being reflected by D-values in or on product. Most pharmaceutical products depress the thermal resistance of micro-organisms relative to their D-values in water, but this is not universally true. Some other materials (e.g., rubber) are known to increase the thermal resistance of microorganisms (this may as likely be due to physical characteristics of heat transference as to biochemical protection). These product effects on thermal resistance can only be determined empirically, and are usually done in the laboratory using thermally resistant bacterial endospores, often spores of *Bacillus stearothermophilus*. The use of *B. stearothermophilus* for this purpose and their frequent use as *biological indicators* (BIs) in bio-validation have contributed to a belief that steam sterilization must be defined in terms of being able to kill this microorganism. It is not; spores of *B. stearothermophilus* are used with steam sterilization because of the convenience of their high resistance to steam sterilization and the unique and distinctive conditions required for their recovery and growth. Indeed, some major companies use *Clostridium sporogenes* or other species of *Bacillus* as reference or indicator organisms for this purpose.

The use of "overkill" specifications is not mandatory. In some instances, there may be pharmaceutical products which are unable to withstand the temperatures or energy inputs of overkill specifications. In these cases, specifications can be developed by calculating SALs of 10^{-6} from data characterizing the number of microorganisms actually contaminating items before sterilization treatment, or from data characterizing the actual numbers and thermal resistances of the contaminating micro-organisms. The question is this—Is this exercise

Table 5.1. D-values of spores of *B. stearothermophilus* on various substrates relative to water (D-value approximately 4 min)

Substrate	*% relative to D-value in water (100%)*
Stainless steel	60
Hydrophobic filter media	90
Silicone tubing	100
Rubber stoppers (various types)	85–150
Polycarbonate	120
Two pharmaceutical products pH 3.4–3.7	15–40
Pharmaceutical product pH 10.5–10.7	115

worth doing or would it be better and simpler to opt for aseptic manufacture of such heat-sensitive products?

Let us consider the number of micro-organisms contaminating pharmaceutical products prior to sterilization. What are the highest and the lowest numbers which could be expected? For sterile parenteral products, the highest tolerable number of microorganisms would be expected to be on the order of 10^2. This is because 10^3 or more per item is likely to begin to incur a risk of pyrogenicity. The lowest number which could be inferred from even an extensive number of zero counts would be one microorganism.

Achievement of a 10^{-6} SAL from an initial bioburden of 10^2 would require eight log reductions. Applying these eight log reductions to an assumed worst case thermal resistance of D_{121}-value in water of 1 min gives a sterilization specification of 121°C for 8 min.

Achievement of a 10^{-6} SAL from an initial bioburden of 1 would require six log reductions. Applying these six log reductions to an assumed worst case thermal resistance of D_{121}-value in water of 1 min gives a sterilization specification of 121°C for 6 min.

The determination of thermal resistances is technically complex and requires special equipment (BIER Vessels). Since it is unlikely that *Bacillus* spp. can be excluded from any survey of microbiological contamination, it is reasonable to assume that spores with D_{121}-values on the order of 0.3 min will be isolated. Using this figure, SALs of 10^{-6} can be calculated at 121°C for 2.4 min for bioburdens of 10^2, and at 121°C for 1.8 min for bioburdens of one microorganism per item.

It is apparent that very brief sterilization specifications (on the order of 2–3 min holding time at 121°C) are obtainable when the microbiological contamination is completely characterized in terms of numbers and thermal resistances. In practice, such limits on hold times could be difficult to control precisely, are probably insignificant in terms of thermal lethality compared with heat-up and cool-down times, and could prove difficult to "sell" to regulators. Without complete thermal characterization of thermal resistances, specifications calculable by the "*bioburden*" approach are hardly significantly shorter than "overkill" specifications. Thus, it probably makes practical sense in most cases to choose only between overkill cycles for thermally resistant products and aseptic manufacture for heat-sensitive products.

Some products may be heat sensitive only above a threshold temperature; for those that can withstand temperatures in the range of 110–118°C but cannot withstand 121°C it is possible to apply the F_0 concept to the principles above and derive equivalent sterilization specifications. As can be seen, if there is a requirement to sterilize at (say) 116°C, there are considerable time savings to be obtained by characterization of the contaminating microorganisms.

Pharmaceutical Products and Materials for Aseptic Manufacture—Sterilizer Parameters

Sterilizer parameters are specific to combinations of product, presentation, and autoclave. They must be established empirically. Heat penetration studies done prior to the performance qualification phase of validation serve the purpose of determining the loading patterns, prevacuums, and temperature and pressure settings, etc. which ensure that the sterilization specification is delivered to the product and that it is delivered uniformly throughout the load.

For instance, a particular proposed loading pattern may never allow for uniform conditions (within specified limits) to be achieved throughout the load. In this case the pattern would have to be changed. Or, in a particular autoclave it may be necessary to set the temperature at 122°C for 121°C to be achieved within the load.

Air removal is particularly important in porous and equipment loads, but is usually of little importance in the sterilization of aqueous pharmaceutical products. Air removal can be important to the specification of new autoclaves—those which are to be designated only for aqueous product sterilization have no need for the pumps and ancillary equipment required to pull deep vacuums.

The involvement of steam in the sterilization of different types of product is an important consideration in understanding and controlling autoclaves.

For aqueous products, steam is solely a means of raising the product to the specified sterilizing temperature; the steam does not come into contact with the contaminating microorganisms. The transfer of heat energy (lethality) to the contaminating microorganisms is from the product itself. To all intents and purposes any suitable form of energy source could be used to raise the temperature of the product. For instance, if ampules of aqueous products were to be sterilized in a hot air oven, the mechanisms of microbial inactivation would still be by coagulation of intracellular proteins. However, heat transfer from hot air is much slower than heat transfer from steam, which is why this is not seen as a practical process. Microwave irradiation could be an alternative means of sterilizing aqueous pharmaceutical products utilizing the same antimicrobial mechanisms as steam; certainly there is evidence that microwave killing patterns are mainly due to heat transfer with very little direct energy being absorbed from the microwaves.

For porous and equipment loads, the steam comes into direct contact with the contaminating microorganisms on the materials being sterilized and there is no intermediary in the transfer of heat. The energy content of steam is defined by its latent heat. If the steam is pure in the sense that it contains neither entrained gas nor moisture, an amount of energy defined by its latent heat at the pressure of the steam will be transferred to the microorganisms by condensation on their surfaces.

There are many potential pitfalls in equipment and porous load sterilization, mainly concerned with air or other non-condensable gas. First, the purity of the steam is important; if it is carrying moisture, or non-condensable gas, it will not contain the same amount of energy as pure steam and its lethality will be less than that predicted for pure steam. Second, any residual air around the contaminating microorganisms may insulate them from contact with the steam and thus reduce the amount of energy (lethality) transferred. In this type of sterilization, steam quality becomes very important and so also do the materials and manner in which the products are contained in steam-permeable wrapping or perforated trays, etc., within the autoclave, and the number and depth of evacuations of the autoclave prior to the temperature- hold phases. Thermal monitoring alone gives little information on the adequacy of the measures put in place to control these complex factors, and it is therefore generally thought essential that some empirical studies be done with BIs as part of process development to ensure that the thermal lethality being imparted by the steam is not being impeded. These development studies may be rolled into bio-validation.

Sterilization of Microbiological Media in the Laboratory

The various suppliers of microbiological media include recommendations for sterilization in their catalog under "Directions for Use," for instance, "sterilize by autoclaving at 121°C for 15–18 min." The question that must be asked is—What do these specifications mean?

Are they intended to apply within the media as are the sterilization specifications for pharmaceutical products and materials for aseptic manufacture? Or are they sterilizer parameters? There may be some indication in some of the older suppliers' manuals which expand their recommendations along the lines of "sterilize by autoclaving at 15 psi (121°C) for 15–18 min." Since pressures of 15 psi are not achievable within media, it is clear that the intention was that the recommendations be applied to sterilizer parameters.

In most cases, it is probably immaterial how these recommendations are interpreted. For media, "*over-cooking*" is bad because of deleterious effects on growth-support characteristics, and "*undercooking*" is generally self-disclosing through evident contamination.

SIP Systems

Systems that are sterilized in-place are often immensely complex. The initial challenges to their sterilization are the removal of air and the elevation of the temperature of the pipework to prevent heat losses and condensation. As such, most work in the development of sterilization specifications for SIP systems is concerned with the heat-up phase. Appropriate questions are: Is the sterilization temperature achieved throughout the system? Where is the slowest location to achieve temperature? Where should the control probe be located?

Often vast amounts of thermal lethality calculated as F_0-values are delivered in these prehold stages of SIP. However, because these temperatures are being achieved in the presence of steam–air mixtures, it is not correct to assume that the biological lethality during the heat-up phase of SIP systems is equivalent to that achieved with pure steam.

The time for which the system must be held at temperature (the sterilization specification) is often relegated to a minor consideration compared with this earlier development work. Typically, it is decided arbitrarily to use, 121°C for 15, 20, or 30 min, with no real scientific basis. Perhaps, a basis parallel to that of the pharmacopeial overkill specifications could be developed. For instance, if the actual maximum number of microorganisms within an SIP system is assumed to be 10^{12} (since this would amount to a few grams of biomass it certainly should be maximal), then 18 log reductions would be required to ensure not more than one chance in a million of a survivor. An overkill cycle of 121°C for 20 min could be proposed by adding two log reductions as a safety factor and assuming each microorganism to have a D_{121}-value of 1 min.

Bio-Validation

The *performance qualification* (PQ) phase of validation follows the development of the sterilization specifications and of the sterilizer parameters which will deliver them. The purpose of PQ in steam sterilization of pharmaceutical products, equipment, laboratory media, and SIP systems is to confirm that the sterilization specification consistently achieves its intended purpose. The process is run using the parameters derived from process development on (usually) three separate occasions and tested for compliance with a variety of predetermined acceptance criteria. As a subset of PQ, the purpose of bio-validation is to confirm that the lethality expected from the process does not significantly deviate from what is expected. Bio-validation is a "test'' of consistency. If the acceptance criteria are not achieved, there may be need for more process development.

In consideration of the extent, thoroughness, and history of the research evidence that microorganisms are inactivated in a regular fashion in response to temperature and time, it is periodically suggested that bio-validation should not be necessary where there is evidence of adequate heat penetration. In practice, however, the expected lethality may not always be achieved. Most frequently, such deviations from ideality occur in equipment and porous load sterilization because of inadequate air removal. Where deviations from ideality occur for aqueous pharmaceutical products, they most likely arise from inadequate knowledge of how the product affects the thermal resistances of microorganisms, but this is best determined in the laboratory at an earlier stage of process development, not at the bio-validation "milestone'' later in the critical path of product introduction.

Acceptance criteria for bio-validation of steam sterilization processes are usually (but not invariably) defined along the following lines:

1. *n*BIs will be placed in the load at locations defined in a drawing.
2. Each BI will contain at least 10^6 viable spores of *B. stearothermophilus*.
3. The load will be exposed to a defined autoclave treatment (the validation cycle).
4. Bio-validation will be considered satisfactory if no viable spores are recovered from the BIs after *x* days of incubation at 55–60°C.

Because this approach is the common practice, there is a widely held belief within the pharmaceutical QA community that the ability to inactivate 10^6 spores of *B. stearothermophilus* is a synonym for achieving an SAL of 10^{-6}. It is not. It is true, however, that inactivation of 10^6 spores of *B. stearothermophilus* with the pharmacopeially approved minimum D-value of 1.5 min in 10–100 replicates guarantees achievement of better than 10^{-6} SALs for worst case bioburdens. However, the converse, i.e., failing to inactivate 10^6 spores of *B. stearothermophilus*, does not necessarily mean that a 10^{-6} SAL has not been achieved.

Another area of confusion is that the USP definition of an overkill specification—"a lethality input of 12D"—can be demonstrated directly in bio-validation. It should be understood that the maximum number of log inactivations of any bacterial population is technically limited to about 9 or 10 D-values. The maximum number of micro-organisms that can be handled as a BI is about 10^7–10^8, the sensitivity of recovery of microorganisms is restricted to more than 10^{-2}.

An indirect demonstration of 12 log inactivations of a microorganism with a D-value of1 min can be achieved by showing inactivation of 10–100 replicate BIs each carrying 10^6 spores with D-values of 1.5 min, or by inactivation of 10–100 replicate BIs each carrying 10^4 spores with D-values of 2 min. Direct demonstration of 12D is technically impossible.

In bio-validation, the spore of B. stearothermophilus is akin to an end-point analytical reagent. For instance, when litmus changes from blue to red at pH levels below 7, it shows only that the pH is not higher than 7. By killing all of 10–100 replicate BIs with 10^6 spores having D-value 1.5 min, all that is proven is that the thermal lethality delivered is not less than an F_0 of 12 min. The PhEur overkill sterilization specification of 121°C for 15 min should meet this requirement easily, and so should any other longer specification at 121°C, or any specification for longer times at lower temperatures taking into account of the F_0 concept.

Numbers and Locations of BIs for Bio-Validation

It is usual for bio-validation to be done with an arbitrary number of BIs between 10 and 100. Both limits are based on practical considerations.

The lower number of BIs is defined in terms of ensuring that bio-validation addresses sufficient parts of the load for confidence that items in all parts of the autoclave are receiving the required lethality. Normal practice is to define this number in terms of placing at least as many BIs as the number of thermal probes used for thermal qualification. It is sensible to place one BI alongside each thermal probe in order to be able to relate thermal data to biological data. In addition to this, some BIs should be placed in other non- probed locations in consideration of the possibility that the leads to the thermal probes may be acting as conduits for air removal or steam penetration, and thus provide falsely high levels of lethality.

More often than not the number of BIs used is about 20–30. Larger numbers up to 100 may be necessary to address very large autoclaves or in thermal mapping studies, but in validation there is little extra statistical confidence to be gained by doing so. In most microbiology QA laboratories, 20–30 BIs can be handled conveniently.

Periodically in the bio-validation of sterilization of porous or equipment "minimum" loads, it is not practical to locate 20 or 30 BIs. For instance, a minimum load may be one cartridge filter, one mop head, or one machine manifold. In such cases, it is important to avoid too much distortion of the statistics of biovalidation. At least five BIs are recommended no matter how difficult it may be to place them.

Choice of BIs

Spores of *B. stearothermophilus* are most commonly used for bio-validation of steam sterilization processes. This is not to say that it is mandatory to use *B. stearohermophilus* nor that it is used exclusively. Other microorganisms, e.g., *sporogenes*, are used by some companies and accepted by the regulatory agencies. Use of *B. subtilis* spores with resistances to steam sterilization in the higher range of that found in natural bioburden has, in recent years, been criticized by European regulatory agencies.

The principles underlying the choice of microorganism used as BIs are quite well known:

1. The microorganisms must have high resistance to the sterilization treatment which they are being used to validate. This does not

mean that they must be the most resistant microorganism known to man. *B. stearothermophilus* has D_{121}-values of 1–4 min according to conditions of culture and the substrate upon which they are mounted. This is higher than most spores of *Bacillus* spp., which tend to have D_{121}-values below 0.5 min.

2. The microorganisms must have stable resistances to the sterilization treatment which they are being used to validate. There are data from commercial suppliers of BIs to show that spores of *B. stearothermophilus* survive and retain stable resistances over long periods of crudely controlled storage.
3. The microorganisms must be easily culturable and preferably be easily identifiable in culture. Very few microorganisms share with *B. stearothermophilus* the ability to grow in simple culture media at 55–60°C.

It is customary to use 10^6 (in practical terms 10^5–10^7) spores per BI. This number is based on custom, practice, and convenience rather than on science. Larger numbers than this are difficult to handle in culture and result in large errors in counting. Smaller numbers reduce the sensitivity of the test. Unfortunately, the widespread use of 10^6 spores for bio-validation has led to a confusion between 10^{-6} SALs and 6 log reductions of *B. stearothermophilus*. More complex decisions surround the choice of substrate within which spores are suspended or on which they are mounted for use as BIs.

The decision tree may be used to help choose the spore substrate used in bio-validation of aqueous pharmaceutical products. To use this decision tree, it is essential to have some knowledge of the effects of product on the resistance of spores; as mentioned before, this requires special equipment and experience.

In all circumstances water is the preferred substrate for bio-validation of aqueous pharmaceutical products. Where water would give deceptive results, it should not be used.

1. If the D_{121}-value of *B. stearothermophilus* is higher in the product than it would be in water (i.e., the product makes the spores more resistant to steam sterilization), then bio-validation must be done with the spores suspended in product. Otherwise falsely favorable results may occur.
2. If the D_{121}-value in product of *B. stearothermophilus* is equal to or less than its D_{121}-value in water, and the sterilization specification is based on overkill, then bio-validation must be done with spores suspended in water. However, if the sterilization specification has

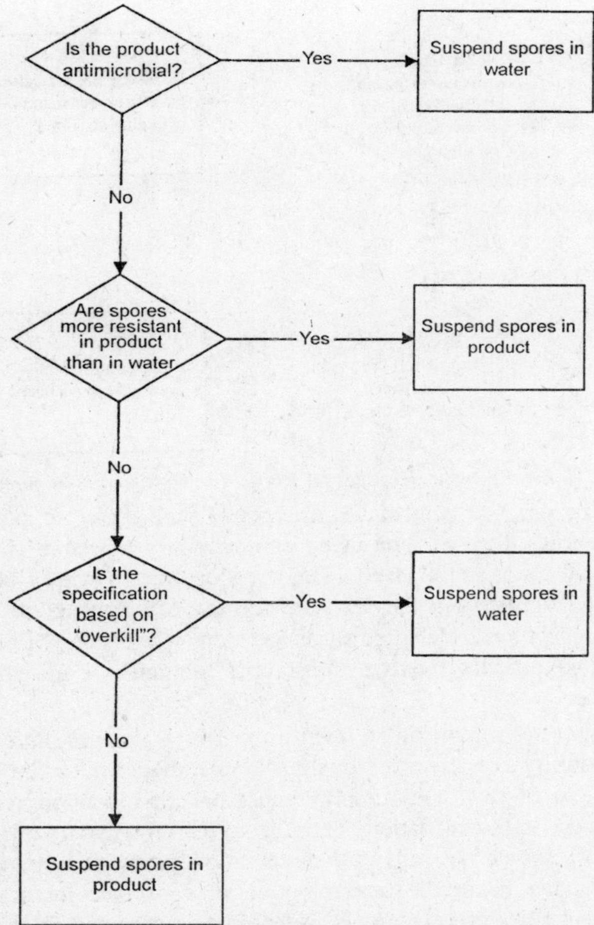

Fig. 5.3. Recommended substrates for BIs used in biovalidation of squeous fluid loads.

been "tailored" specifically to the resistance of microorganisms in the product, then bio-validation must be done with the spores suspended in product. Otherwise falsely unfavorable results may occur.

Under no circumstances must spores be suspended in product if that product is sporicidal.

For other materials (for instance, equipment and supplies for aseptic manufacture) where the mechanisms of inactivation rely on direct contact between the steam and the item being sterilized and therefore

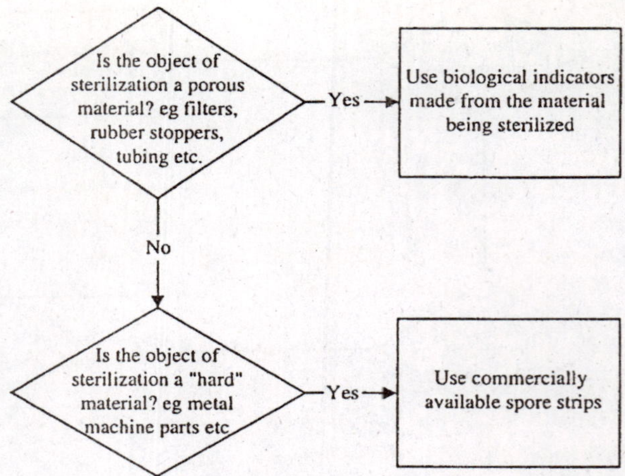

Fig. 5.4. Recommended substrates for BIs used in biovalidation of porous loads.

where air removal is matter of importance, the choice of substrate for BIs generally lies between using commercially available paper spore strips and the material itself. The decision tree in 4 may be helpful. Regulatory pressure is currently toward use of inoculated product, but commercially available spore strips are more convenient. "Tailor-made" inoculated product requires substantial amounts of microbiological expertise.

Use of commercially available BIs transfers much of the responsibility for assuring quality in manufacture to the supplier. Regardless of this, their quality must be controlled on receipt and prior to use in bio-validation. There is no reason why any microbiology laboratory should not verify the numbers of spores per commercial BI. On the other hand, the determination of resistance requires special equipment and expertise and is probably best accepted on the basis of the supplier's certification. If this is done, the user of the BIs is responsible for knowing what the certified measures of resistance mean, how they were determined (on the strips, in aqueous suspension, or in or on something else), and that they were determined correctly and in compliance with applicable standards and legislation.

Validation Cycle

Bio-validation is usually done against a sterilization specification which delivers less lethality than the lower limit of lethality allowed by the sterilization specification defined for the material being sterilized. It is clearly intellectually flawed to choose to validate something different to that which is ever to be used in practice. So what is the reasoning behind this practice?

Sterilization specifications in the "hold" period are presented in terms of temperature and time with upper and lower tolerances set around them. The lower specification limits are critical to sterilization.

Time is generally easily controllable to quite high levels of accuracy and precision: A steam valve allows steam to enter the autoclave until the hold temperature is reached; the valve is then closed and the process is controlled by a timer which, at the end of the specified hold period, sends a control signal to activate the exhaust valves and cooling sequences. The hold time is usually specified in terms of whole minutes—well within the accuracy and precision of all but the most inappropriate of timers.

Temperature is less easy to control precisely. The temperature in the hold period in autoclaves is generally maintained by modulating valves which open to allow steam entry when the temperature (or pressure, because these valves are more often than not controlled through pressure transducers) begins to drop toward the critical lower limit of the specification. It is generally not possible to control an autoclave to run through a complete hold period at the lower limit of its temperature specification. However, even quite apparently trivial errors in temperature above or below the limit can make significant differences in the amount of lethality delivered. For instance, at a nominal temperature of 121°C, an error of 1 K can increase or decrease the amount of lethality by 25%.

Bio-validation cycles are therefore designed to ensure that no more lethality is delivered than that specified by the lower limits of lethality of the sterilization specification used in routine practice. Ideally this is done by reducing the time of the hold period, but sometimes, when quite short cycles are being used, it may also be necessary to reduce the temperature set point on the autoclave as well. The risk in all of this is that the bio-validation cycle is used as a justification for the release of sterilized items when specifications are not complied with under atypical production conditions. This idea should not be entertained.

Acceptance Criteria

Bio-validation is a limit test, which at best produces quantal data. Each BI should be tested separately for survivors or absence of survivors. The acceptance criterion should be that there are no survivors. Detection of survivors in all exposed BIs is clearly unacceptable; such a result would be obtainable if the BIs had never been near a sterilizer. Data showing survival on some but not all of the BIs may be valuable in process development (particularly in the development of sterilization specifications and sterilizer parameters for porous and equipment loads), but would likely raise issues at regulatory inspection. The inference taken from having some survival could be that each item in the autoclave load is not being exposed to the same treatment. For initial validation of a newly developed process, complete inactivation of all BIs should not be difficult to achieve.

The range of D_{121}-values acceptable to USP for spores of *B. stearothermophilus* allowed as BIs for use in steam sterilization is 1.5–3.0 min. An overkill sterilization or bio-validation specification delivering an F_0 of 15 min would deliver 10 log inactivations or, if there were 10^6 spores per BI, one chance in one million of finding a survivor on any one BI, one chance in one thousand of finding a survivor in 10 BIs, one chance in one hundred of finding a survivor in 100 BIs, and so on. In other words, there are pretty long odds against failing the acceptance criteria.

However, spores of *B. stearothermophilus* may have D_{121}-values of 3 min. In such a case there would be practically no chance of meeting the acceptance criteria of killing 10^6 spores with an F_0 of 15 min. What are the implications of this?

On the face of it, the implication is that this sterilization specification/sterilizer parameters combination is invalid. However, remember that this same sterilization specification/sterilizer parameter combination would have been valid if the BIs used had D_{121}-values of 1.5 min. In practical terms the pharmacopeial specification for thermal resistance in BIs has been set naively. Many companies purchasing spores of *B. stearothermophilus*, either for preparing BIs or as commercial strips, order against their own specifications which include upper D_{121}-value limits (in water) of around 2 or 2.5 min. The author of this paper has published recommendations for determining biovalidation cycles appropriate to challenge numbers and D_{121}-values of the spores available, but in the long run it is far more convenient and easier to justify compliance to regulatory agencies when there is bio-validation to show that 10^6 spores have been inactivated in 10–100 replicates.

Requalification

Periodically it is wise to repeat bio-validation. Changes do occur in autoclaves and no change control procedure, no matter how rigorously implemented, is infallible. The purpose of requalification is to determine if any unforeseen change has arisen to affect the sterility assurance provided to the items being sterilized.

It is important for requalification that the numbers, resistances, and substrates for the BIs are closely similar to those used in the initial validation. For the same reasons as resistance variation within BIs as discussed before, it is possible if these factors are not well controlled to emerge from requalification with either a false confidence in the security of the process (use of BIs which are less resistant than those used in initial validation), or with the incorrect opinion that the process has failed (use of BIs which are more resistant than those used in initial validation).

Biological requalification is usually done following significant process changes or on an annual frequency. The establishment of a frequency should, in principle, be based on business risk; in fact, however, with well- designed and controlled autoclaves, the risk to the business of extending the interval beyond 1 yr is probably more one of incurring regulatory criticism at inspection than of releasing non-sterile products to market.

Bio-Validation of Laboratory Autoclaves Used for Sterilizing Microbiological Media

It is not difficult to argue that the effectiveness of sterilizing microbiological media is self-disclosing and should not therefore merit bio-validation. The pertinence of bio-validation to the qualification of laboratory autoclaves is more to do with having confidence in the sterility of media before or after it is used in the laboratory (and risk false positive results if the media is not sterile), or take it into (say) aseptic manufacturing areas for environmental monitoring (and if it is non-sterile contaminate areas and products which may otherwise have been secure). Many regulatory bodies see bio-validation of laboratory autoclaves as mandatory.

Bio-Validation of Steam-in-Place (SIP) Systems

SIP systems range from very small systems (say, a mixing tank) where all parts may reach temperature within 2 or 3 min, to absolutely massive arrangements of vessels, valves, and pipework in which the expulsion of air, the drainage of condensate, and the attainment of the sterilization specification temperature at the "slowest point" can take 20 or 30 min.

Bio-validation is essential. The placement of BIs is largely a matter of judgment. Certainly the "slowest point" to reach the sterilization specification temperature must be challenged. Certainly vent filters and low points where condensate could accumulate must be challenged. Thereafter, it is a regulatory expectation that there should be sufficient BIs placed to give coverage to the whole system, which in effect may mean placing BIs in locations which, due to the heat-up time of the system, have been exposed to the sterilization specification temperature for two, three, or four times longer than the "slowest point." Undue confidence should not be taken from favorable results from these locations.

6

PROTEIN FARMING

The demand of the pharmaceutical industry for large quantities of mammalian proteins has led to the development of heterologous expression systems for the production of proteins and peptides of varying complexity. The host organisms used range from bacteria to eukaryotic systems such as yeast, insect, mammalian and plant cell cultures and transgenic animals and plants. Bacteria can produce relatively high levels of foreign protein but develop insoluble inclusion bodies and offer only limited post-translational modification. In contrast, eukaryotic expression systems are able to glycosylate proteins and carry out post-translational processing, although different post-translational effects can result in the formation of products that are not identical in all respects to the native protein. The cost of protein production using different host organisms and expression systems varies widely. The complexity of the protein, its end use, the scale of production and the degree of similarity required between the transgenic and native proteins are important factors to consider when determining the type of production system to apply.

Plant-based production systems are now being used commercially for the synthesis of foreign proteins. Post-translational modification in plant cells is similar to that carried out by animal cells; plant cells are also able to fold multimeric proteins correctly. The sites of glycosylation on plant-produced mammalian proteins are the same as on the native protein; however, processing of *N*-linked glycans in the secretory pathway of plant cells results in a more diverse array of glycoforms than is produced in animal expression systems. Glycoprotein activity is retained in plant-derived mammalian proteins.

Agricultural production of foreign proteins in crop plants can deliver large quantities of product at low cost. This is possible even if protein expression levels are low relative to other heterologous systems, as agriculture is a cheap technology and production targets can be achieved using additional plantings or cropping larger areas of land. The ability to scale-up economically and to utilize production and extraction technologies already developed for the food industry significantly reduces the cost of high-volume protein production using agriculture compared with alternative methods such as cell culture. The cost of agricultural production is also reduced substantially if the foreign protein does not require purification from the plant biomass, as is the case for edible vaccines.

Despite the advantages of agriculture, and even if the product can be shown to have appropriate biological activity, agricultural production of foreign proteins may not provide adequate assurance of product safety and quality. For example, foreign proteins produced in the field are subject to contamination with pesticides, herbicides and mycotoxins; field-grown plants also experience variable weather conditions, non-uniform soil compositions and infestation by pests and diseases. Any or all of these factors may result in unpredictable product yield and quality. The inability to control production conditions could mean that whole-plant systems fail to comply with good manufacturing practice in many countries around the world, particularly if the transgenic protein is to be used as a therapeutic. Issues of environmental crop safety also arise, especially if the foreign protein is toxic, e.g. to soil microorganisms or to wild-life capable of consuming the plants. In situations where the plant-derived product has suitable activity but where contamination, inconsistent quality and/or regulatory issues prohibit the use of agricultural methods, large-scale plant tissue culture offers an alternative route for foreign protein production. This is particularly soif the volume of protein required is low.

As well as overcoming many of the inherent problems associated with agriculture, plant tissue culture also offers a number of advantages over conventional animal cell culture methods currently being applied to produce biopharmaceutical proteins commercially. As plant culture media are relatively simple in composition and do not contain proteins, the cost of the process raw materials is reduced and protein recovery from the medium is easier and cheaper compared with animal cell culture. In addition, as most plant pathogens are unable to infect humans, the risk of pathogenic infections being transferred from the cell culture via the product is also substantially reduced.

Production of Foreign Proteins Using Plant Tissue Culture

In this review, we focus on the use of plant tissue culture to produce foreign proteins that have direct commercial or medical applications. The development of large-scale plant tissue culture systems for the production of biopharmaceutical proteins requires efficient, high-level expression of stable, biologically active products. To minimize the cost of protein recovery and purification, it is preferable that the expression system releases the product in a form that can be harvested from the culture medium. In addition, the relevant bioprocessing issues associated with bioreactor culture of plant cells and tissues must be addressed.

Extensive research has been carried out into the molecular aspects of foreign protein production in whole plants to enhance the yield, quality and stability of the product and to facilitate protein separation and purification from the biomass. In contrast, comparatively little research has been undertaken to investigate the specific issues associated with producing foreign proteins in plant tissue culture. Many of the problems that need to be addressed, such as low protein yields, are similar *in vivo* and *in vitro*. However, the differences between these systems means that different solutions may be required.

Some developments that have proven useful for enhancing foreign protein yields in whole plants, such as chloroplast and organ-specific expression systems, have little or no practical application in tissue culture. On the other hand, although not yet demonstrated *in vitro*, the use of viral vectors to increase protein production, and novel approaches for simplifying product purification such as targeting of proteins to oil bodies, may have significant implications for the production of foreign proteins in tissue culture systems.

Several of these proteins, e.g. antibodies, interleukins, erythropoietin, human granulocyte-macrophage colony stimulating factor (hGM-CSF) and hepatitis B antigen, have pharmaceutical or therapeutic uses and would be suitable for further commercial development if the production levels could be increased. However, to date, few plant cell or organ cultures have been shown to accumulate or secrete foreign proteins at concentrations sufficient for commercial viability.

Tobacco (*Nicotiana tabacum*) has been the host species in most studies of foreign protein expression in plant tissue culture. Rice (*Oryza sativa*) has also been used by several groups. Accumulation of hGM-CSF was found to be significantly higher in rice suspensions with an inducible promoter than in tobacco suspensions with constitutive transgene expression. The predominance of tobacco-based expression systems in tissue culture studies differs from the situation with whole plants, where advantages associated with producing foreign proteins in edible species or in storage organs such as seeds have resulted in a variety of plant species being transformed.

Most research into *in vitro* foreign protein production has been undertaken using cell suspensions. However, other forms of plant tissue culture such as hairy roots and shooty teratomas have also been tested in a number of studies. The characteristics of different types of plant tissue culture and their utility for large-scale foreign protein production are outlined in the following sections.

Suspended Cell Cultures

Plant cell suspensions comprise small clumps of dedifferentiated plant cells in liquid nutrient medium. Dedifferentiation of the cells occurs under the influence of plant growth regulators, which must be provided in the medium to promote rapid growth and maintain the culture morphology. Transgenic cell suspensions can be developed from callus initiated using explants from transformed plants; alternatively, wild-type suspensions may be transformed directly using *Agrobacterium tumefaciens*-mediated transfection or biolistic delivery of plasmid DNA into the cells.

Plant suspensions are being used to produce an increasing number of foreign proteins. These include complete antibodies, antibody fragments, hGM-CSF, interleukin-2, interleukin-4 and interleukin-12, erythropoietin, hepatitis B surface antigen, α_1-antitrypsin, human lysozyme and carrot invertase. Although, in some instances, stable production of foreign proteins has been found to occur over extended periods, suspension cultures are subject to various types of genetic instability through the effects of somaclonal variation. Significant reductions in the yield of foreign proteins over time, possibly caused by genetic instability, have been reported in plant cell suspensions. In some cases, cell lines with stable production characteristics were isolated by screening and selection from a large number of cultures.

Hairy Root Cultures

Hairy roots are neoplastic roots produced by transformation of plant cells with *Agrobacterium rhizogenes*. When cultured in liquid medium, hairy roots often exhibit rapid growth relative to untransformed roots. Hairy roots can be propagated indefinitely in liquid medium and retain their morphological integrity and stability in the absence of exogenous plant growth regulators. Hairy root cultures have been found to have significantly greater long-term stability than suspended plant cells for the production of foreign proteins.

Transgenic hairy root cultures can be initiated by infecting transgene-containing plants or explants with *A. rhizogenes*. Using this approach, it is relatively easy to generate hairy roots expressing multiple foreign genes, as plants containing multiple transgenes (produced by sexually crossing transgenic plants carrying single transgenes) may be used for hairy root initiation. For example, transgenic tobacco plants developed by crossing antibody-heavy-chain-expressing plants with antibody-light-chain-expressing plants were used subsequently to generate hairy roots capable of synthesising complete IgG_1 antibody. Transgene-expressing hairy roots can also be obtained by performing root initiation and transformation at the same time using genetically-modified *A. rhizogenes* with the transgene inserted into plasmid constructs. Alternatively, established hairy root cultures can be induced to produce foreign proteins by direct *A. tumefaciens*-mediated transformation.

Shooty Teratoma Cultures

Shooty teratomas are a form of differentiated organ culture produced by transformation of plants with particular strains of *Agrobacterium tumefaciens*. Foreign protein production in transgenic shooty teratomas has been reported by only one group. In this system, shooty teratomas of tobacco were used to produce an IgG_1 antibody. Antibody yields in the teratoma cultures were lower than in suspended cell and hairy root cultures. The growth characteristics of shooty teratomas were also not conducive to liquid culture as the shoots tended to callus and were very susceptible to hyperhydricity (vitrification).

Scale-up Considerations for Different Forms of Plant Tissue Culture

Several studies of foreign protein production have been carried out using plant cell suspensions or hairy roots in bioreactors. Bioprocess development for the large-scale culture of suspended plant cells is relatively well established, as this type of culture has been examined extensively for the production of plant secondary metabolites such as paclitaxel (taxol), ginseng and shikonin. Accordingly, if suspension cultures suitable for the commercial production of foreign proteins were developed, the basic technology for large-scale operations is already available. Research into bioreactor systems for more complex forms of tissue culture such as roots and shoots is not as well developed. The principal reactor types trialed for large-scale hairy root culture have been reviewed by Giri and Lakshmi Narasu. Difficulties associated with providing a low-shear environment while maintaining adequate mixing and oxygen transfer present significant problems for the scale-up of root reactors. However, despite these engineering challenges, if advantages such as enhanced culture stability are associated with hairy roots compared with suspended cells, further technical development of root cultures for foreign protein production would be worthwhile.

Strategies for Improving Foreign Protein Accumulation and Product Recovery in Plant Tissue Culture

As indicated by the protein accumulation levels, it is currently possible using plant tissue culture to achieve moderate levels of foreign protein expression in some systems. To take advantage of the cost benefits associated with recovering products from the culture medium rather than from homogenized biomass, expression systems for protein secretion have also been developed. Yet, relative to the production levels attained in animal cell cultures, foreign protein concentrations in plant cultures are typically very low. This is a major hurdle preventing plant systems being utilized more widely for commercial protein production. Therefore, a key research objective has been to increase the accumulation of active foreign proteins in plant tissue culture. Although the reasons for the low yields in plant systems are not yet fully understood, two different approaches have been taken to increase product levels *in vitro*. These are: (i) to increase the level of gene expression in the cells by altering the transgene constructs and methods of expression, and (ii) to increase the retention and stability of foreign protein in the cultures after the protein is produced. Efforts have also been made to enhance the availability of product in the culture medium to facilitate subsequent recovery and purification.

Expression Systems

Compared with whole plants, there has been limited development of foreign protein expression systems specifically for use in tissue culture. Some modifications of expression constructs have resulted in improved protein accumulation or have allowed simplified protein recovery. However, in general, modified expression systems have been tested only in a restricted number of cases and have not resulted in the large increases in product yield required for plant cultures to compete with other foreign protein production vehicles. Transient expression techniques, for example using viral vectors, that have been developed for use in whole plants have not yet been applied in plant tissue culture.

Modifications to existing expression constructs

Several molecular strategies have been successful in increasing foreign protein production in cultured plant cells. These include using promoters for inducible expression, optimizing codon usage and adding the KDEL sequence to ensure protein retention in the endoplasmic reticulum. Application of different promoters has been the most common approach to the modification of expression constructs.

Most of the promoters used in plant tissue culture have been based on the constitutive *cauliflower mosaic virus* (CaMV) 35S promoter. In contrast, inducible promoters have the advantage of allowing foreign proteins to be expressed at a time that is most conducive to protein accumulation and stability. Although a considerable number of inducible promoters has been developed and used in plant culture applications, the only one to be applied thus far for the production of biopharmaceutical proteins is the rice α-amylase promoter. This promoter controls the production of an α-amylase isozyme that is one of the most abundant proteins secreted from cultured rice cells after sucrose starvation. The rice α-amylase promoter has been used for expression of hGM-CSF, α_1-anti-trypsin and human lysozyme.

Alterations to the proteins and pre-proteins expressed by cultured plant cells have been used to facilitate product recovery. A leader sequence is required for foreign protein secretion from plant cells into the apoplast and then into the culture medium. As indicated, plant, mammalian and viral sequences have been employed to achieve the entry of transgenic proteins into the bulk-flow pathway in plant cultures.

To facilitate product recovery and purification, molecular tags may be added to foreign proteins. Attachment of a functional His_6 tag to a secreted therapeutic protein, hGM-CSF, has been examined in tobacco suspension cultures. The His_6 tag consisted of six histidine residues attached to the protein terminus and gave the protein the ability to bind strongly to metal ions. The presence of the His_6 tag allowed the specific removal of product from the culture medium using iminodiacetic acid metal affinity resin.

Transient expression using viral vectors

Genetically modified viral vectors have been applied in many whole-plant systems for the production of therapeutic proteins and epitope vaccines. Foreign proteins produced using viral vectors can be in the form of free cytosolic proteins or fusions to viral proteins. Viral expression systems exploit the ability of viruses to propagate rapidly and achieve high concentrations in plant tissues. For example, *tobacco mosaic virus* (TMV) can accumulate in infected tobacco leaves to levels greater than 60 mg g^{-1} dry weight and produce amounts of TMV coat protein accounting for 10–40% of the total protein content of the leaves. Provided the movement proteins on recombinant viruses remain functional, viral vectors are able to spread throughout the entire plant from a single infection point via the plasmodesmata between individual cells and the vascular system. Therefore, in principle, when foreign protein is co-expressed with the virus, large amounts of product can be formed.

To date, application of transgenic viruses in whole plants has not resulted in the production of foreign proteins to the same high levels as viral proteins from non-transgenic virus infections. This is probably because the genetic construct carried by the virus interferes to some extent with the normal folding, packaging, transmission or replication processes. Nevertheless, foreign protein yields achieved using viral vectors can be substantial. For example, transgenic viruses with coat protein fusions have been reported to accumulate to levels of 1–3 mg g^{-1} of plant tissue.

A vector that facilitates high-level protein expression in plant tissue culture, particularly a transient expression system that could be applied to existing wild-type cultures, would be advantageous for *in vitro* foreign protein production. However, such a system has not yet been developed. The success of this approach depends in part on whether appropriate levels of viral infection, replication and transmission can be established within tissue culture systems.

Previous work has shown that mechanical inoculation techniques, for example, rubbing cells with abrasive powder or vibrating cell suspensions in a vortex mixer, can be used to infect plant cell suspensions with viruses. Other methods that have been tested include microinjection of viruses into plant cells and inoculation of callus by pricking the tissues with needles dipped in a virus suspension. Viral infection has been reported to occur to some extent even without special mechanical treatment of cultured plant cells. It is thought that viral agents are able to enter cells via the plasmodesmata observed to be present in dedifferentiated cultures. Infection of suspended cells was found to be most successful when friable cell clumps were freshly dispersed from callus into liquid medium containing the virus. The reason given for this was that the protoplasmic connections between the cells were broken in this procedure so that the plasmodesmata were exposed allowing viral entry into the cells. Additional non-intentional injury may also occur to cells in agitated culture, thus providing other routes for virus infection.

In previous work, levels of viral accumulation in plant cell suspensions have been significantly lower than those achieved in whole plants. For example, suspended tobacco cells have been reported to accumulate only one-thirtieth to one-fortieth the concentration of TMV attainable in tobacco leaves. In other experiments, maximum TMV coat-protein levels of only about 250 $\mu g\ g^{-1}$ fresh weight were measured in tobacco cell suspensions. Even though plant cells in suspension tend to aggregate so that individual cells in clumps are connected by plasmodesmata, the spread of virus between infected and uninfected cells that are not in direct contact is likely to be very limited. Therefore, compared with the recombinant systems already available for tissue culture applications, suspended plant cells may offer no significant advantage for improving foreign protein yields using virus-based expression. However, other forms of plant tissue in which the cells are in close and constant contact with each other, such as differentiated organs, may prove feasible hosts for high-level viral expression of foreign proteins *in vitro*.

We have examined the characteristics of virus infection of hairy roots to determine if root cultures would be suitable for foreign protein production using viral vectors. Extensive cell-to-cell contact occurs in hairy roots and some viral transport may also be possible through the vascular tissue. TMV was produced in significant quantities in *Nicotiana benthamiana* hairy roots, mostly during the period of active root growth. The concentration of virus in replicate cultures was found to vary considerably. The average concentration of virus between days 21 and 36 was approximately 2 $mg\ g^{-1}$ dry weight; however, levels of virus in individual cultures were as high as 5 $mg\ g^{-1}$. These results demonstrate the potential of hairy roots for the propagation of plant viruses. Further work is underway to test the production of co-expressed foreign proteins in hairy root cultures using a modified TMV vector.

Secretion of Foreign Proteins

Proteins produced in plant cells can remain within the cell or are secreted into the apoplast via the bulk transport (secretory) pathway. In whole plants, because levels of protein accumulated intracellularly,

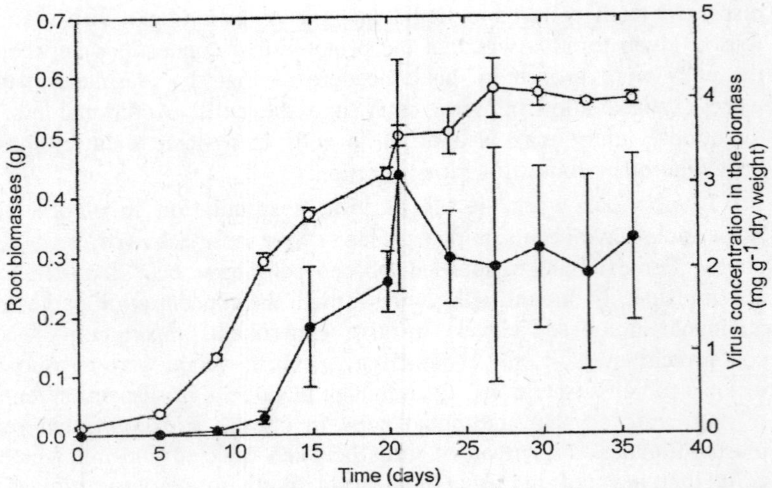

Fig. 6.1. Root growth (o) and accumulation of tobacco mosaic virus (TMV) (•) in hairy roots of N. benthamiana.

e.g. using the KDEL sequence to ensure retention in the endoplasmic reticulum, are often higher than when the product is secreted, foreign proteins are generally not directed for secretion. However, as protein purification from plant biomass is potentially much more difficult and expensive than protein recovery from culture medium, protein secretion is considered an advantage in tissue culture systems. For economic harvesting from the medium, the protein should be stable once secreted and should accumulate to high levels in the extracellular environment.

Secretion of foreign proteins into the medium requires that the protein molecules move through the cell walls. The pores in plant cell walls are thought to allow passage of globular proteins of maximum size around 20 kDa; however, a small number of wider pores may serve as channels for relatively slow permeation of larger molecules. Foreign proteins with molecular weights significantly greater than 20 kDa have been recovered in substantial quantities from plant culture media. In other cases, despite having signal sequences that allow the protein to reach and traverse the plasma membrane, recombinant proteins such as erythropoietin and IgG-2b/κ antibody remain associated with the plant cell wall and fail to be released from the biomass. These results suggest that protein composition and structure may affect the extracellular availability of secreted foreign proteins.

The presence of foreign protein in the medium of plant cultures does not necessarily mean that all or even most of the product can be recovered from the medium. In many expression systems where an appropriate signal sequence has been used, considerable amounts of foreign protein remain within the plant cells and/or tissues. For example, in a comparison of IgG_1 antibody production in tobacco cell suspension and hairy root cultures, a maximum of 72% of the total antibody was found in the medium of the suspension cultures whereas only 26% was found in the medium of the hairy root cultures. This result could indicate that secretion and/or transport across the cell wall was slower in the hairy roots; alternatively, it could indicate poorer stability of the secreted protein in the hairy root medium. If foreign proteins are to be purified from the medium, improved secretion and extracellular product stability are desirable.

Foreign Protein Stability

There is considerable evidence that foreign proteins are subject to a significant degree of degradation and instability in plant expression systems, both inside and outside of the cells.

Stability inside the cells

Foreign protein fragments often appear in addition to the intact protein in western blots of extracts from transgenic plants and plant cells. This phenomenon is not confined to plant tissue cultures or particular host species, and occurs in seeds as well as vegetative tissues. A detailed investigation of IgG_1 antibody fragments in tobacco cell suspension and hairy root cultures has been carried out by Sharp and Doran. Although various explanations have been offered for the presence of foreign protein fragments, such as protease release during sample homogenization, the presence of assembly intermediates, and variations in the extent of protein glycosylation, in the case of IgG_1 antibody these explanations could not adequately account for all the observed molecular properties of the fragments. Instead, with the aid of a range of affinity probes, glycosylation and secretion inhibitors and glycan-reactive agents, proteolytic degradation in the apoplasm of tissues such as hairy roots, and between the endoplasmic reticulum and Golgi apparatus in both hairy roots and suspended cells, was identified as the most likely mechanism of fragment formation.

Stability outside the cells

The simplicity of plant culture media is considered an advantage for foreign protein production in tissue culture systems. However, as a mixture of salts and sugar containing several heavy metals but negligible protein (except for any protein secreted from the plant cells), plant culture medium provides an environment for foreign proteins that is very different from the physiological conditions inside the cells.

The effect of medium composition on the concentration of IgG_1 antibody in solution is illustrated. In this experiment, antibody was added at a concentration of 1.0 mg L^{-1} to fresh, sterile media in shake flasks. The flasks were then incubated on an orbital shaker at 25°C and the antibody concentration was measured as a function of time using an *enzyme-linked immunosorbent assay* (ELISA). There is a significant difference between the extent of antibody retention in Murashige and Skoog (MS) plant culture medium and in media designed to support the growth of animal cells. After 7 hours, about 80% of the added antibody was retained in Dulbecco's minimal essential medium (DMEM) containing 10% fetal bovine serum and about 70% was present in serum-free Ex-cell 302 medium. In contrast, in MS medium, less than 10% of the added antibody could be detected after only 1.5 h. These results indicate that fresh, sterile plant culture medium does not support the retention and stability of proteins in solution.

There have been many reports from several groups that plant culture medium is not conducive to protein stability, and that the retention of secreted proteins in culture media can be very poor. The mechanisms responsible for protein loss from plant culture media are not completely

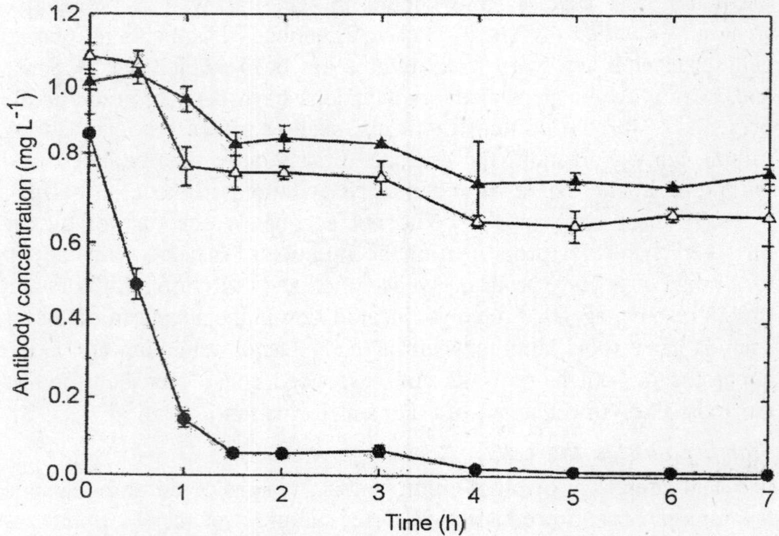

Fig. 6.2. Stability of IgG_1 monoclonal antibody added to sterile plant and animal cell culture media.

understood; however, current indications are that multiple factors may be involved. Processes that have been proposed to affect foreign proteins in plant media include protein degradation due to protease activity, protein instability due to defined or undefined conditions or components in the medium, surface adsorption of proteins onto the culture vessel and protein aggregation or insolubility.

As declining levels of foreign protein in plant tissue culture have been associated in a number of studies with an increase in the concentration of extracellular proteases, minimizing protease levels in the medium and/or reducing the susceptibility of heterologous proteins to protease degradation have been investigated as methods for improving foreign protein accumulation. Several approaches have been tested to achieve this objective with varying levels of success. These include adding the broad-spectrum protease inhibitor bacitracin to the medium of cell suspension and hairy root cultures, adjusting the osmolarity of the medium to minimize cell disruption and protease release, adding gelatin as a possible alternative substrate for protease activity, reducing protease accumulation by using inducible promoters to allow separation of the growth and production phases and using host species such as rice that are considered to secrete lower levels of proteases than the more commonly applied tobacco.

Medium additives

Protein stabilizing agents such as polyvinylpyrrolidone (PVP), gelatin, *bovine serum albumin* (BSA) and salt (NaCl) have been demonstrated to improve the retention of several types of foreign protein in plant tissue culture media. The precise mode of action of these additives in protecting proteins is unclear; however, they may prevent protein aggregation, conformational change and/or adsorption onto the internal surfaces of the holding vessel. The appropriate stabilizing polymer must be identified for each foreign protein production system, as their effects appear to vary depending on the specific culture and its protein product. The addition of biopolymers, particularly proteins such as BSA, to tissue culture media has the potential to complicate downstream processing operations for product recovery. However, when the resulting increase in protein yield is large, the additional cost of product purification may be acceptable.

Several medium additives other than the protein stabilizing agents mentioned above have also been tested to improve foreign protein accumulation in plant tissue cultures. These include dimethylsulfoxide (DMSO), polyethylene glycol, nitrate, amino acids, heamin, gibberellic acid and glutamine. The mechanisms by which these components might affect intra- and extracellular foreign protein levels include improving protein expression and synthesis, increasing protein secretion, reducing the extent of intracellular protein degradation and improving protein stability in the medium.

The results of empirical studies carried out to test the effects of medium additives on foreign protein accumulation in plant tissue culture are summarized below.

Polyvinylpyrrolidone (PVP)

PVP is a metabolically inert, water-soluble polymer with excellent protein stabilizing properties. As an example of the beneficial effect of PVP on foreign protein accumulation in plant tissue culture, data for growth and IgG_1 antibody levels in transgenic tobacco hairy root cultures with and without PVP are shown. In these experiments, PVP with a relative molecular mass of 360,000 was added to the medium at a concentration of 1.5 g L^{-1}. PVP had no significant effect on root growth; in a number of studies, PVP at concentrations up to 3 g L^{-1} has been found to have little effect on growth of plant cell and organ cultures. The primary effect of PVP was a substantial increase in the amount of foreign protein in the culture medium; the maximum level of antibody in the medium with PVP was about four-fold greater than that without PVP. As indicated, on average throughout the culture period the effect of PVP on antibody levels in the root biomass was relatively small. Similar results have also been reported for PVP-treated plant cell suspensions producing foreign protein.

The addition of PVP 360,000 at a concentration of 0.75 g L^{-1} has been reported to yield a 35-fold increase in the level of extracellular foreign protein in suspended plant cell cultures. The effectiveness of PVP in stabilizing secreted proteins depends on both the polymer molecular weight and its concentration. Low-molecular-weight (10,000 and 40,000) PVP was found to be less effective than PVP 360,000. Increasing concentrations of PVP 360,000 up to 1.0 g L^{-1} improved antibody accumulation in hairy root culture medium; however, above this concentration there was no further increase in antibody levels. Addition of PVP after extracellular foreign protein levels had decreased during plant suspension culture did not result in a recovery of the protein. Although PVP has proven successful as a foreign protein stabilizer in several systems and has yielded substantial increases in product concentrations as discussed above, it has been found relatively ineffective in other plant tissue cultures producing foreign proteins.

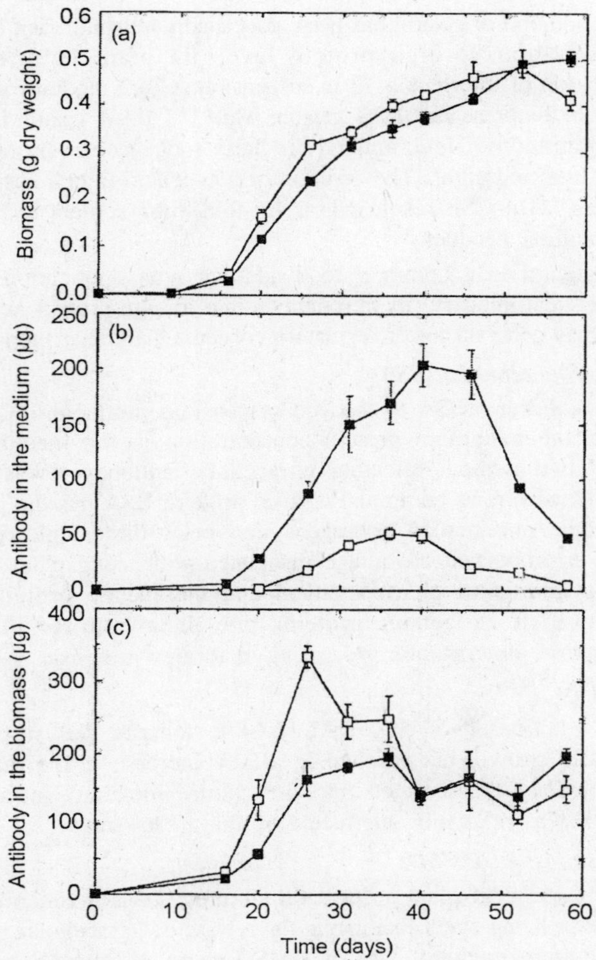

Fig. 6.3. Effect of PVP on (a) growth; (b) antibody in the medium; and (c) antibody in the biomass, for N. tobacum hairy roots expressing IgG_1 antibody.

Gelatin

Gelatin has been shown to enhance foreign protein levels in the medium of trans- genic plant tissue cultures. However, growth in both suspended cell and hairy root cultures was reduced, with the negative effect increasing with gelatin concentration. It has been suggested that gelatin could act either as a protein stabilizing agent or as an alternative substrate for protease activity.

The addition of gelatin has been associated with significant increases in extracellular foreign protein levels in plant cultures. The concentration of interleukin-12 in suspension culture medium increased four-fold in the presence of 2% gelatin, while 0.1–0.9% gelatin increased the concentration of IgG_1 antibody in hairy root medium by factors of between four and eight. The addition of 5% gelatin to cell suspensions expressing hGM-CSF resulted in a 4.6-fold improvement in the yield of extracellular product.

As gelatin is a common food additive with applications in the pharmaceutical industry, its introduction into foreign protein production systems may generate fewer regulatory concerns than other biopolymers.

Bovine serum albumin (BSA)

The addition ofBSA to tobacco suspensions producing hGM-CSF increased the maximum protein concentration in the medium by a factor of two without affecting intracellular antibody levels or cell growth. However, as an animal-derived protein, BSA has the potential to introduce mammalian pathogens into plant tissue cultures, thus negating an important advantage associated with using plant systems for the synthesis of pharmaceutical and therapeutic proteins. The introduction of exogenous proteins into plant cultures may also complicate the downstream processing of foreign proteins.

Salt (NaCl)

The addition of 50–100 mM NaCl to tobacco cell suspensions reduced cell growth but resulted in a 50% increase in the maximum level of hGM-CSF secreted into the culture medium. Intracellular hGM-CSF was relatively unaffected by this treatment.

Dimethylsulfoxide (DMSO)

When added to plant suspension cultures, DMSO functions as a cell permeabilizing agent facilitating the release of intracellular products into the culture medium. Use of DMSO as an effector for antibody secretion and/or a protein-stabilizing agent has been tested using tobacco cell suspensions and hairy roots. The addition of DMSO to suspended cells expressing an antibody heavy chain resulted in a significant increase in both intracellular and extracellular product levels. Whereas untreated cultures accumulated a maximum of 150 $\mu g\ L^{-1}$ of total antibody, treated cultures accumulated 210 $\mu g\ L^{-1}$ in the presence of 2.8% DMSO and 430 $\mu g\ L^{-1}$ in the presence of 4% DMSO. In contrast, when DMSO at a concentration of 2.8% was added to transgenic hairy roots at various times during the culture, there was no significant effect on IgG_1 antibody accumulation in either the biomass or medium.

Nitrate

Supplementation of Gamborg's B5 medium with 0.1% KNO_3 (in addition to the 0.25% KNO_3 usually present in B5 medium) resulted in a significant increase in antibody levels in tobacco hairy root cultures. The antibody concentration in the medium increased 2.8-fold and the total antibody accumulation increased by 90% relative to cultures without added nitrate.

Amino acids

Transgenic tobacco suspensions supplemented with a cocktail of essential and non-essential amino acids produced three-fold more IgG-2b/κ antibody than untreated cultures. The amino acids were added 10 h prior to harvesting of the cultures in the presence of 1 mM $CaCl_2$ to facilitate amino acid uptake.

Other additives

Further additives have been tested for their influence on foreign protein accumulation in plant tissue cultures. These include heamin ($C_{34}H_{32}O_4N_4FeCl$), an effective inhibitor of ubiquitin-dependent proteolysis, the plant hormone gibberellic acid, which has been shown to enhance the activity of the endoplasmic reticulum and secretory pathway in plant cells, and the amino acid glutamine. However, these compounds had relatively little effect on foreign protein levels. Addition of polyethylene glycol to plant culture medium also did not significantly improve the accumulation of foreign protein in the systems tested.

Medium properties

Modifying the properties of plant culture media, including increasing the osmolarity, reducing the effective concentration of selected heavy metals and altering the pH, has resulted in enhanced foreign protein accumulation or stability in some systems.

Osmolarity

Hyperosmolar medium is known to be advantageous for the production of proteins such as antibodies in animal cell cultures. Likewise, raising the osmolarity of MS medium using mannitol was found to improve the accumulation of foreign protein in both the biomass and medium in plant cell suspensions. When IgG_1-secreting tobacco cell suspensions were grown in hyperosmolar medium, maximum antibody levels in the medium increased by up to 2.4-fold compared with standard MS medium. This result was possibly due to greater amounts of antibody being produced or secreted from the cells, as IgG_1 antibody added to fresh, sterile hyperosmolar medium did not show increased stability compared with unmodified medium. The influence of medium osmolarity on foreign protein accumulation and activity was also demonstrated by Terashima *et al.* using rice suspensions. In this system, α_1-antitrypsin was expressed using an inducible promoter that allowed protein production to occur when the sucrose content of the medium was low. Sucrose starvation was achieved by replacing the plant growth medium with sugar-free production medium. However, the removal of sucrose decreased the osmolarity of the medium and the activity of the α_1-antitrypsin produced was relatively low. Increasing the osmolarity of the production medium using mannitol or N-(2-hydroxyethyl) piperazine-N'-(2-ethanesulfonic acid) (HEPES)/NaCl resulted in a significant improvement in α_1-antitrypsin activity.

Heavy metal concentration

The heavy metals copper, manganese, cobalt and zinc were omitted individually and in combination from MS and B5 media to determine the effect on antibody stability in solution. When IgG_1 antibody was added to these modified media in experiments, only the B5 medium without Mn showed a significant improvement in antibody retention relative to normal culture media. Nevertheless, protein losses were considerable as only about 30% of the added antibody could be detected in the Mn-free medium after about 5 h. The beneficial effect of removing Mn was lost when all four heavy metals, Cu, Mn, Co and Zn, were omitted simultaneously. The reason for these results is unclear. Addition of the metal chelating agent ethylenediaminetetraacetate (EDTA) had a negligible effect on antibody retention in both MS and B5 media.

pH

Solution pH has a strong influence on the structure, conformation and solubility of globular proteins. However, when IgG_1 antibody was added to fresh, sterile B5 medium adjusted to pH values between 4.0 and 8.0, there was no significant difference in the rate at which the antibody disappeared from the different solutions. In other work, antibody-expressing tobacco hairy roots were cultured in B5 medium with initial pH between 3.0 and 11.0. Root growth was affected severely at the lowest and highest pH values and total antibody levels declined as the initial pH was increased above 5.0–6.0.

Bioprocess Developments

The development of novel bioreactors and bioreactor operating strategies has the potential to improve the performance of cell culture systems. Some progress has been made in this area to enhance foreign protein production in plant tissue culture.

Product recovery from the medium

Continuous or periodic harvesting of secreted foreign proteins from plant culture media could be used to increase product yields by removing active protein before it can be degraded or otherwise lost from the culture.

Shake-flask-based affinity-chromatography bioreactors have been used for simultaneous production and purification of foreign proteins *in vitro*. Heavy chain antibody secreted by suspended plant cells was recovered by continuously recycling the culture medium through a column containing Protein G resin. The harvested antibody was eluted from the column at the end of the culture period. The extent to which the antibody bound to the resin depended on the recirculation flow rate and the column pH. Although extracellular antibody concentrations were similar in the affinity-chromatography and control flasks, protein harvesting increased the total production of secreted protein more than eight-fold. In similar experiments, hGM-CSF with a His_6 tag was harvested semi-continuously by recycling the medium through a metal affinity column. Cell growth was reduced in this system; however, the total amount of hGM-CSF accumulated was more than twice that obtained without product removal.

In other work, periodic foreign protein recovery from cell suspension and hairy root cultures was achieved using hydroxyapatite resin. Beginning 13 days after inoculation of the hairy roots and 7 days after inoculation of the suspensions, medium was periodically separated from the biomass, exposed to the resin and then returned to the cultures. Recovered IgG_1 antibody was eluted from the resin and the total amount of antibody produced by the cultures determined by ELISA. Even though growth was unaffected by the hydroxyapatite, the overall effect of this process on antibody accumulation was relatively small, with maximum total antibody levels only 20–21% higher with periodic harvesting than without. Considering the rapidity with which antibodies may be lost from plant culture media, continuous rather than periodic removal of product may be required to achieve more substantial improvements in yield.

Oxygen transfer and dissolved oxygen concentration

Adequate aeration of plant tissue cultures is crucial for achieving maximum production of foreign proteins. Poor oxygen transfer has been found to limit cell growth and reduce antibody heavy chain production in genetically modified tobacco cell suspensions in shake flasks. However, although raising the air flow rate in a stirred 5-L bioreactor increased foreign protein levels to a certain extent, excessive aeration resulted in foaming and reduced both growth and antibody production.

In transgenic hairy root cultures, the concentration of IgG_1 antibody in the biomass increased by 52% relative to the levels obtained under air when the roots were cultured at a dissolved oxygen tension of 150% air saturation. Compared with root cultures grown at 50% air saturation, which is a realistic operating dissolved oxygen tension in poorly mixed, large-scale root reactors, oxygen enrichment to 150% air saturation improved total antibody accumulation 2.9-fold.

Plant tissue culture offers an alternative to agriculture as a plant-based system for producing low-to-medium volumes of foreign proteins. Secretion of foreign proteins into the culture medium has significant potential benefits for reducing the cost and complexity of product recovery and purification. A range of molecular, culture and bio-processing techniques has been employed to enhance the accumulation and stability of foreign proteins in plant cell and organ cultures. Further improvements in product yield are required to raise the economic competitiveness of plant tissue culture as a viable commercial method for the controlled, safe and reliable manufacture of pharmaceutical and therapeutic proteins.

7

CONSERVING FARM PRODUCTS

Microbial spoilage of pharmaceutical products has been known for many years. Spoilage may result in the deterioration of the product due to loss of potency or to the initiation of an infection in the user. Sterile pharmaceutical products (single dose or multiple dose forms) require the addition of an antimicrobial preservative when they have been manufactured under aseptic conditions from presterilized ingredients. Where the products are subject to a terminal sterilization process, only the multiple dose category requires the addition of an antimicrobial agent. In the latter instance, the preservative is added to protect the product and end user against the consequences of microbial entry during use. Chemical antimicrobial agents are thus added to all multidose sterile formulations and to aqueous and aqueous-based non-sterile pharmaceuticals. Their function is to reduce the microbial load to a level, which is safe for the designated use of the product, and to maintain the numbers of viable microorganisms at or below that value for the storage and use life of the product. Preservatives must, therefore, be stable within the formulation for the shelf life of the product and be capable of dealing with all the abuses made to it by the consumer and user (i.e., contamination during use, incorrect storage etc).

For multidose sterile products, the preservative must be capable of reestablishing sterility between each use, whereas for a non-sterile topical cosmetic the function of the preservative might simply be to prevent growth. The associated toxicity of preservatives often limits the concentrations at which they can be employed; thus, lower concentrations are generally employed for opthalmic products and injectables. In choosing a preservative the likely capacity required, the rate of killing desired, and the ingredients and pH of the formulation must be borne in mind.

The BP (1988) test for the "Efficacy of Preservatives in Pharmaceuticals" makes it quite clear that a preservative must in the first instance *kill* microorganisms and only accepts later that it will prevent growth. However, it does insist that there must be no outgrowth of the test organism, that is, the preservative possesses sufficient residual capacity to inhibit the growth of any survivors.

Definition of Terms

Disinfectants, antiseptics, and preservatives are chemicals that have the ability to destroy or inhibit the growth of microorganisms, and are used for this purpose. These and other terms commonly employed are defined as follows:

1. *Disinfectants*: Chemical agents or formulations that are too irritant or toxic on body surfaces, but are used to reduce the level of microorganisms from the surface of inanimate objects to one that is safe for a defined purpose.
2. *Antiseptics*: Chemical agents or formulations that can be used as an antimicrobial agents on body surfaces.
3. *Preservatives*: Chemical agents or formulations that are capable of reducing the number of viable microorganisms within an object or field to a level that is safe for its designated use and will maintain the numbers of viable microorganisms at or below a level for the use/shelf-life of the product.
4. *Bacteriostasis*: A state in which the growth of microorganisms is halted or inhibited.
5. *Bactericide*: A chemical antimicrobial agent that reduces the viability of a population of microorganisms exposed to it. This term is meaningless without specifying the concentration range over which this effect is obtained; such concentration ranges will vary between different species of microorganisms.
6. *Bacteriostat*: A chemical antimicrobial agent that can prevent the growth of microorganisms within an otherwise nutritious environment. This term is meaningless without specifying the concentration at which this effect is achieved. Bacteriostatic concentrations do vary between different species of microorganisms.

It should be noted that terms such as bactericide and bacteriostat should be discouraged; in the USP and EP, the term "*antimicrobial agent*" has replaced these terms.

Preservative Ideals

At present there is no perfect preservative, and all materials are a compromise of a number of often contrary properties. The following are the properties of an ideal preservative compound and need to be considered when choosing a preservative.

1. *Definable in chemical terms*: Many of the existing preservatives, such as the quaternary ammonium compounds, are mixtures of various homologues. Often the activity obtained is a function of the mixture composition. Unless it is possible to define and control mixture composition, the performance of the agents will be variable, even if they conform to a pharmacopoeial specification.
2. *Broad spectrum of activity*: The compounds must possess a broad spectrum of antimicrobial activity against all species of micro-organisms and also toward bacterial endospores. In practice, the only compounds that meet this requirement are formaldehyde, gluteraldehyde, hypochlorite, and ethylene oxide. All these compounds are highly irritant at sterilizing concentrations to be used in pharmaceutical products. Formaldehyde is, however, used at low concentrations in some shampoos; in these cases contact with the skin is short-lived and irritancy minimal. Agents such as quaternary ammonium compounds, phenolics, and the parabens group possess good activity against gram-positive bacteria but little or no activity toward spores. Certain gram-negative organisms such a *Pseudomonas aeruginosa* are virtually resistant to these agents. Generally, antifungal activity is difficult to obtain. Combinations of preservatives are sometimes employed to widen the spectrum of activity to include molds, bacteria, yeasts, and endospores.
3. *Effectiveness*: The compounds must be effective over a wide range of pH in order to be effective in all formulations. In practice, compounds are generally more active at either acid or alkaline pH. Thus, the pH of a formulation determines the types of preservative suitable for inclusion.
4. *Stablility*: The compounds must be stable to light and elevated temperatures for the expected shelf life of the product. The effects of pH upon stability should be minimal. In this respect it is worth noting that the preservative Bronopol is stable only in the dark and at an acid pH. Under alkaline conditions or in the light it rapidly decomposes to give formaldehyde at concentrations that would be ineffective as a preservative. Instability to light can be protected against the packaging in a light-proof container. Storage

tests must be performed on all formulations to ensure that adequate levels of preservative remain at the end of then expected shelf life of the product.

5. *Solubility*: Preservatives should ideally be used at concentrations much lower than that of the main constituents of the formulation. Their solubility ought to be such that it is possible to add them as a concentrated solution and where there is no danger of creating a saturated solution.
6. *Aesthetics*: Preservatives should have no perceptible odor, color, or taste, which might affect the aesthetic qualities of the final product. This can be of crucial importance for a cosmetic product but is less important for medical ones.
7. *Volatility*: Preservatives should be non-volatile. Thus, chloroform is not an ideal preservative as it is lost from the formulation each time it is exposed to air.
8. *Product incompatibility*: Preservatives should not be incompatible with any of the likely excipients within the product formulation. This would include incompatibilities with the container material and also the active ingredients. In practice this is very difficult to achieve.
9. *Toxicity*: At the concentrations employed, the preservative should be non-irritant, not cause hypersensitivity reactions, and be non-toxic. In this respect, the site of application is critical. Relatively few compounds are approved for use in opthalmic products due to their high sensitivity towards xenobiotics. Also, compounds safe for use on intact skin might be hazardous for inclusion in parenteral products.
10. *Solubility in oil*: Preservatives must not be too oil soluble as this can produce problems in two- and three-phase systems where the preservative accumulates in the oil and micellar phases and is unavailable for antimicrobial action in the biological (aqueous) phase. It is worth noting that the oil: water partition coefficient can alter as a function of pH and also as a function of the nature of the oil.
11. *Cost*: The preservative must be cost-effective in the context of the overall product positioning (i.e., cost of goods ratio).

Although a very large number of antimicrobial compounds have been examined for their suitability for preservation of pharmaceuticals, only a few (<20) are currently used in the majority of pharmaceutical products. Moreover, in many cases, the "least unsuitable" rather than the "optimum" preservative is selected for a particular product.

Dynamics of Preservation

The critical lethal parameters of the effectiveness of antimicrobial agents are the concentrations of the agents employed, the time of exposure, the types and numbers of organisms exposed, the pH of the environment or formulation, and the temperature of application. An understanding of the effects of these conditions upon the killing process and of the terms that describe such dependence is important to rationalize the use of preservatives in different situations and formulations.

Time of Contact

When exposed to lethal concentrations of a chemical antimicrobial agent, the viability of a bacterial population decreases exponentially with time. Such pseudo-first-order kinetics can be described by the equation

$$\frac{S}{S_0} = Ae^{-kt} \qquad \text{...(1)}$$

$$\log\frac{S}{S_0} = kt + \log A \qquad \text{...(2)}$$

where k is the rate constant, S is the number of surviving organisms at time t, and S_0 is the initial number of viable organisms within the population.

D-value is the time for 90% kill. That is,

$$\log S - \log S_0 = -1 \qquad \text{...(3)}$$

and

$$D\text{-value} = 1/k \qquad \text{...(4)}$$

In practice, such obedience to first-order kinetics is rare for chemical inactivation. More commonly, deviations from first-order kinetics are obtained, when there is either an initial lag in the rate of killing or when the rate of killing decreases with time of exposure. The former is commonly observed when the concentrations of preservative are minimally bactericidal and reflects the time required for injuries caused to the bacterial cells to assume lethal proportions. Such plateauing therefore decreases as concentrations of agent employed become greater and are inapparent at very high concentrations. The second effect reflects the presence of resistant organisms within the population, capacity effects on the action of the preservative, or failure

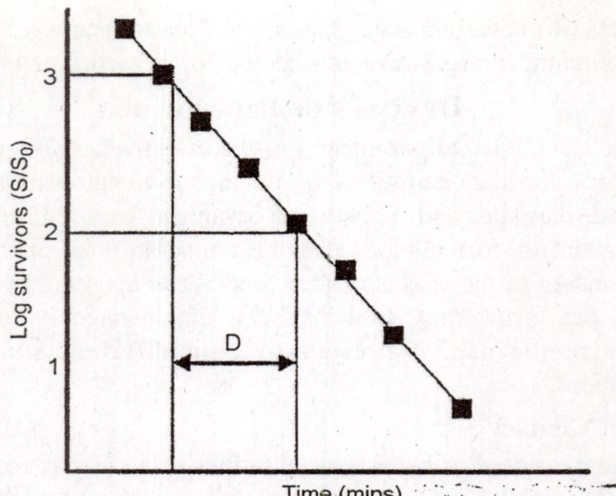

Fig. 7.1. Time survival kinetics for the inactivation of a bacterial population by treatment with a chemical agent.

of the neutralizer. In such cases commonly observed with quaternary ammonium compounds and biguanides, the levels of kill at the plateau value increase with increasing concentration of preservative. From these relationships, an estimate of the initial rate of killing (k) or the D-value (actual time to effect a 90% reduction in viability) can be made, but their predictive value must be questioned.

Approximation to first-order kinetics implies that the sensitivity toward preservatives of a population of microorganisms is normally distributed. Decreases in the rate of killing with time often results

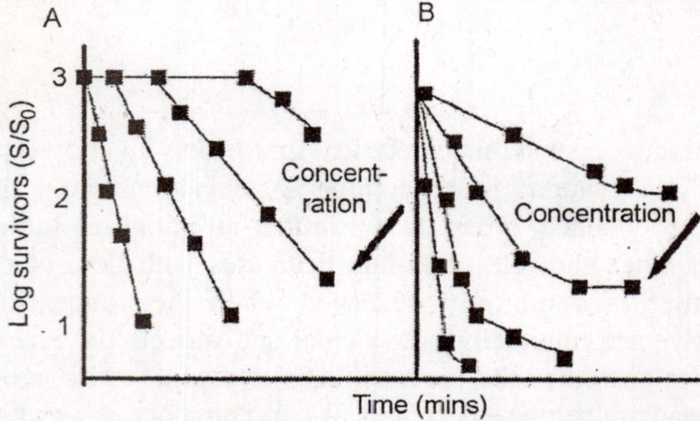

Fig. 7.2. Deviations from first-order kinetics of the inactivation of microbial populations by treatment with chemical agents due to (A) an initial lag in the rate of killing or (B) a decrease in the rate of killing with time of exposure.

from adsorptive loss of thepreservative onto and into the killed cells, causing a decrease in the available concentration of the preservative for killing the surviving organisms. The susceptibility of preservatives to such adsorptive losses is described as its capacity.

Exponential decreases in viability with time mean that total elimination of all living cells cannot be guaranteed. Rather, knowing the initial size of the population and its susceptibility, the probability of having achieved sterility might be assigned to particular inactivation processes. Such assignments assume, often wrongly, that first-order kinetics are universally applicable.

Concentration of Preservatives

The efficacy of an antimicrobial agent varies as a direct function of its concentration. At very low concentrations the antimicrobial effect might be negligible, at intermediate concentrations the effects might be growth-inhibitory, whereas only at relatively high concentrations might they be bactericidal. Often different mechanisms of action are involved to bring about these different effects. For example, chlorhexidene inhibits growth through direct effects upon respiratory enzymes and ATP synthesis, but kills through causing gross permeability changes to the bacterial cell envelope. The minimum growth inhibitory concentration (MIC) of an agent gives little or no information about its bactericidal activity, but might indicate weaknesses in the overall antimicrobial spectrum of activity. Minimum lethal concentrations can range from 2 $\times$ MIC – 20 $\times$ MIC, dependent upon the antimicrobial substance and target organism. In the bactericidal range, the concentration exponent (η) describes the dependence of activity on concentration. This can be obtained as the slope of a graph of log *D* or *k* versus log concentration. Although these values vary slightly among organisms, they are fairly characteristic of compound groups and also often relate to mammalian toxicity. Thus, compounds with high concentration exponents more readily lose their biological activity when diluted upon parenteral administration than those with low exponents.

The concentration exponent gives no indication as to the level of activity of a particular agent, It only indicates the degree of dependence of activity upon concentration. If the $\log_{10}$ of a death time is plotted against the $\log_{10}$ of the concentration, a straight line is usually obtained, the slope of which is the concentration exponent. Given the *D*-value for a single concentration and concentration exponent, it is possible to calculate the activity at any other concentration. As different anti-microbial mechanisms are often involved ingrowth inhibition and killing,

Table 7.1. Some examples of preservative concentration exponents

Compound	*Exponent*
Organic mercurials	0.9–1.0
Iodine	0.9–1.0
Formaldehyde	1.0–1.1
Benzalkonium chloride	0.8–2.5
Bronopol	0.7–0.9
Parabens	2.5–3.0
Phenolics	4.0–9.9
Aliphatic alcohols	6.0–12.7

the use of concentration exponents to extrapolate MIC data to bactericidal situations should be avoided.

$$(\eta) = \frac{(\log D \text{ at } C_2 - \log D \text{ at } C_1)}{\log C_1 - \log C_2} \quad \ldots(5)$$

The value of the concentration exponent also gives an indication of the susceptibility of the agent to inactivation by dilution and/or losses in activity through adsorption of the agent on organic debris or killed bacterial cells. The loss of an antimicrobial agent from solution in this manner relates to concentration in solution and absorptivity rather than to the inherent antimicrobial activity of the agent. The activity and concentration exponents of four hypothetical agents are described. Agents A & B and C & D have similar concentration

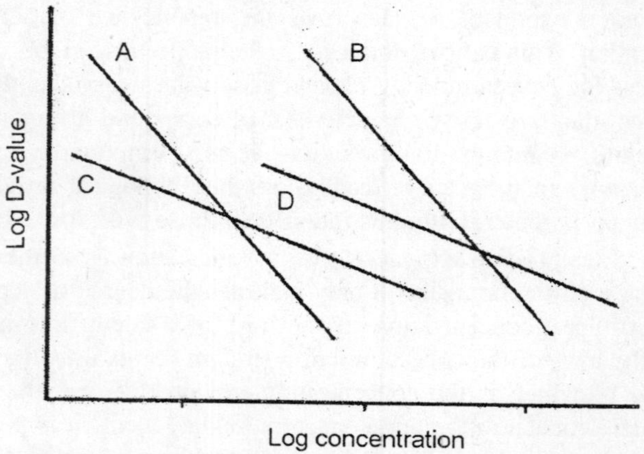

Fig. 7.3. Comparisons of the activity and concentration exponents four hypothetical preservatives.

exponents but different activities, whereas A & C and B & D have similar activities yet different concentration exponents.

If isotoxic concentrations of each agent (equivalent *D*-value) are employed, and undergo a uniform drop in concentration through adsorption, the compounds with the higher activity will undergo the greatest loss of activity (A & C > B & D). For compounds of similar activity, the one with the higher concentration exponent will have the greatest activity loss (A > C and B > D). This can be generalized in the statement that capacity is inversely related to both the concentration exponent and activity. Preservatives used in sterile products generally have high concentration exponents and high activity. Such considerations are also helpful in determining inactivation conditions for the preservatives prior to sterility and preservative effectiveness tests.

Temperature Effects

The rate of killing of an antimicrobial agent is directly dependent on the temperature of the interaction. This dependence is described by the temperature coefficient (Q or θ), which can be determined either from the slope of the graph relating the *D*-value or *K* and temperature, or from Eq. (6).

$$\text{Temp: Coefficient } [\theta_{t1-t_2}] = \frac{D\text{ - value } t_2(D_2)}{D\text{ - value } t_1(D_1)} = \frac{kt_1}{kt_2} \qquad \ldots(6)$$

The subscripts denote the temperature range over which the change in activity is defined. This is commonly given as the coefficient per 10°C rise (i.e., θ_{10} or Q_{10}) but is sometimes quoted for 1°C and 25°C changes.

Although the temperature coefficients vary a little between organisms, they are characteristic for particular preservative groups. A high θ_{10} value renders a compound suitable for application at elevated temperatures but unsuitable for the preservation of formulations that might be stored over a wide temperature range.

It should be noted that the temperature will also affect the degree of ionization in some circumstances; therefore, the concentration exponent for some compounds will be apparently affected by temperature.

Effect of pH

The degree of ionization of acidic and basic antimicrobial agents depends on pH. Some compounds are active only in the unionized state (e.g., phenolics) whereas others are preferentially active as either

Table 7.2. Temperature coefficients (10°C) for some commonly used preservatives

Compound	θ_{10}
Aliphatic alcohols	30–50
Phenolics	3–5
Formaldehyde	1.5
Chlorocresol	6.0
Organic mercurials	1.0

the anion or cation. It therefore follows that the activity of a particular concentration of an agent will be enhanced at a pH that favors the formation of the active species. Thus, cationic antibacterials such as acridines and quaternary ammonium compounds are more active under alkaline conditions. Conversely, phenols and benzoic acid are more active in an acid medium. Chlorbutol is less active above pH 5 and unstable above pH 6. Phenylmercuric nitrate is only active at above pH 6 whereas thiomersal is more active under acid conditions. The sporicidal activity of glutaraldehyde is considerably enhanced under alkaline conditions whereas hypochlorites are virtually ineffective at above pH 8.

Knowledge of the pKa of a preservative allows the calculation of the fraction undissociated at any particular pH and hence (via the concentration exponent) the level of activity at that pH.

Fraction of undissociated biocide

$$= \frac{1}{1} + \text{antilog}\,(\text{pH} - \text{pKa}) \qquad \ldots(7)$$

The pH also affects activity by altering the net surface charge on the microorganisms. Thus, as pH is decreased, the surface becomes less negative and adsorption and thereby activity of cationic agents is reduced. These effects are greater for gram-negative than gram-positive bacteria.

Presence of Interfering Substances

In most instances, before an antibacterial agent can act on a cell, they must first combine. The presence of other materials, commonly referred to as organic matter, may reduce the effect of such an agent by adsorbing or inactivating it and thus reducing the amount available for combining with target cells. Extraneous matter may be able to form a protective coat around the cell, thereby preventing the penetration of the active agent to its site of action.

Preservative Interactions

Until the late 1930s, cosmetic and pharmaceutical formulations were stabilized with soaps. Although preservatives were often required, spoilage was only rarely a problem. With the advent of new materials for emulsion stabilization, preservation of the systems, or the lack of it, became a major problem. Nowadays we are aware of apparent incompatibilities between the new excipients and the established preservative agents and of the many interactions that can lead to preservative failure. No all-embracing set of rules have emerged to enable the "correct" preservative to be chosen for each formulation; only general guidelines exist.

The simplest form of interaction is one of adsorption/ adsorption of the preservative from the formulation into and onto container and excipients. In this respect, it is convenient to divide these into colloidal excipients, suspensions, metal ions, and polymers.

Colloids

1. *Gums*. Tragacanth, acacia, and plant polysaccharides such as agar and alginates are generally anionic in nature and will adsorb cationic preservatives such as chlorbutol, benzalkonium chloride, substituted phenols, phenylmercuric acetate, and merthiolates. Generally the adsorptive process increases with the increase in concentration of the gum or polysaccharide. Colloidal solutions of tragacanth and acacia, but not agar or alginates, can often be preserved with cationic agents by increasing the concentration of the preservative to saturate the colloid. For preservatives such as parabens, the problem is exacerbated at alkaline pH when the preservative becomes fully ionized.
2. *Starches*. These bind substituted phenols but this can be overcome by increasing the concentration of the phenolic. Gelatin shows a slight interaction with the parabens group.
3. *Proteins*. These interact both with cationic and anionic preservatives depending on to the pH of the formulation and the net charge of the protein.
4. *Non-ionic surfactants*. Interactions with a range of preservatives are unpredictable. Cationic and anionic surfactants interact and are contraindicated with anionic and cationic preservatives respectively.
5. *Polyethylene glycols*: These interact particularly with the phenolic group of preservatives. The interaction increases with increase in

molecular weight of the polymer. With increasing temperature, the extent of binding increases due to thermal desolving of the polymer and the creation of further binding sites.

Suspending Agents

These include clays, bentonite, kaolin, and talcs. They are generally anionic in nature and interact with most cationic agents. Inactivation depends on the pH of the system. Preservatives are only loosely associated with clays and therefore, desorb on dilution. The extent of these interactions are difficult to predict.

Container Materials

Nylon strongly adsorbs phenols and sorbic acid by covalent linkage. The degree of interaction depends on the pH, nature of the solvent, temperature, contact time, surface area, preservative concentration, and thickness of the nylon.

Acrylates and polyethylene affect most preservatives, including organic mercurials, phenolics, and benzoic acids. Leaching of hydroxyl ions from glass raises the pH and indirectly affects preservative activity. In this respect, glass must pass a "*limit-test*" for alkalinity.

Changing the container of a product from glass to plastic can alter its preservation. The container is as much a part of the formulation as is the product that is placed in it. Equally, adsorption and absorption into rubber closures can be significant, as exemplified by the phenolics.

Table 7.3. Loss of preservative through absorption into rubber caps

Preservative	*Initial concentration (% w/v)*	*Final concentration (% w/v)*	*Percentage loss*
Phenol	0.5	0.39	22
Cresol	0.3	0.21	30
Chlorocresol	0.1	0.04	60
Phenyl mercuric nitrate	0.001	0.00005	95

Preservation of Two-phase Systems

When preservatives are added to two-phase oil–water systems, depending upon the hydrophilicity of the agent, they partition between the two phases. Bacteria will only grow within the aqueous phase or at the oil- water interface. Thus, preservative lost to the oil is unavailable for antibacterial action. Hence the concentration in the water phase is reduced, affecting the rate of kill, but the capacity of the system is unaffected as preservative lost by adsorption onto organic debris and bacteria is replaced from that held in the oil.

The distribution between the oil and water phases is easily predicted from the oil-water partition coefficient, K_W^0

$$K_W^0 = \frac{\text{Concentration in oil}}{\text{Concentration in water}} \quad ...(8)$$

By knowing the concentration of preservative required for the necessary biological effect in the aqueous phase (C_w), the oil: water phase ratio (θ) and the K_W^0 for the system, C, the concentration needed inthe formulation, may be calculated from the expression

$$C_W = \frac{C(\theta + 1)}{K_W^{0\ \theta+1}} \quad ...(9)$$

It is also worth noting that altering the nature of the oil can affect partition coefficients quite dramatically, for example, the differences between vegetable and mineral oils.

Table 7.4. Partition data for various preservatives

	K_W^0	
Preservative	*Mineral oil*	*Vegetable oil*
Chlorocresol	0.5	11.7
Methyl parabens	0.02	7.5
Propyl parabens	0.5	80.0
Butyl parabens	3.0	280.0
Cetrimide USP	<1.0	<1.0
Bronopol	0.043	0.11
Phenonip	>1.0	>1.0
Phenyl mercuric nutrate	<1.0	<1.0

If the preservative favors the oil, increasing the oil: water ratio dramatically decreases available concentration; if it favors the aqueous phase, the reverse occurs and can cause solubility problems. Unfortunately, both temperature and pH also affect K_W^0, in addition to the effects of pH directly on the activity of the preservative.

Preservation of Three-phase Systems

Most oil-water systems contain surfactants as emulgens in order to provide a stable emulsion. Some preservatives may associate directly with the surfactants whereas others may partition into the micellar phase of the surfactant, treating it as a third phase for partitioning.

These effects can be calculated to some extent, where C_W depends not only on K_W^0 but also on surfactant concentration. Initially it is assumed that the surfactant micelles form part of the aqueous phase. Thus, if K_W^0 is the oil : aqueous phase partition coefficient and CA the total aqueous concentration, then, as in Eq. (9),

$$C_A = \frac{C(\theta+1)}{K_A^{0\theta+1}} \quad \text{...(10)}$$

Now

$$C_W = \frac{C_A}{R}$$

where $R = C_A/C_W$ that is $C_A = RC_W$

As $K_W^0 = C_A / C_W$ and $C_0 = K_W^0 C_W$

substitution in Eq. (9) gives

$$C_A^W = \frac{C(\theta+1)}{C_0/C_A^{\theta-1}} \quad \text{...(11)}$$

and substituting once again for C_0,

$$C_A = \frac{C(\theta+1)}{(K_W^0 \frac{C_W}{C_A\theta})+1} \quad \text{...(12)}$$

Substituting in Eq. (12) for $C_A = RC_W$ gives

$$RC_W = \frac{C(\theta+1)}{(K_W^0\, C_W/RC_W\theta)+1} \quad \text{...(13)}$$

which then reduces to

$$C_W = \frac{C(\theta+1)}{[K_W^0\theta + R]} \quad \text{...(14)}$$

Eq. (14) allows the active concentration in the water phase to be calculated from C, the phase ratio (θ), K_W^0 and R. This is useful as K_A^0 varies with the concentration of the surfactant. R can be found from a graph relating C_A/C_W and surfactant concentration, where the slope of the line is k.

$$R = 1 + k \text{ [surfactant concentration]}$$

All preservative manufacturers provide values of k for a range of surfactants.

Toxicity and Safety Considerations

There are certain limitations on the use of antimicrobial preservatives based on their toxicity and potential side effects. The

British Pharmacopeia recommends that the maximum quantity of an injection containing an antimicrobial preservative to be administered at a single occasion is 15 ml. The USP, on the other hand, recommends a maximum of 5 ml. These maximum volumes must be considered when toxicity data and concentrations of the agents included in the formulations are considered. The final concentration of any substance employed is a balance between its potential toxicity and antimicrobial activity. Additionally, because meninges are readily irritated, the inclusion of antimicrobial preservatives in any formulation that might gain access to the cerebrospinal fluid is precluded. Intracardiac and intraarterial injections must also not contain an antimicrobial preservative.

It is not surprising that preservatives sometimes prove toxic to humans. To assess the toxicity of any substance, it is necessary to know its acute and cumulative toxicity together with its "no effects levels" and also to be aware of its mutagenic, teratogenic, and carcinogenic potential. In addition, its local actions as well as its liability to cause hypersensitization might be important. Any new preservative must be subjected to a battery of such mandatory tests to provide these data as must a new application of an established biocide.

Acute Toxicity

Preservatives, for the most part, are used at very low concentrations and the question of their acute toxicity rarely arises. The margin of safety is not always very high, however.

Subacute and Chronic Toxicity

When substances are given repeatedly, manifestations of toxicity may differ, particularly if the compound is given at a rate faster than what the individual can detoxify or excrete. Some preservatives contain mercury and this may accumulate in the body.

Reproductive Toxicity

As part of the general safety testing of all preservatives, consideration is give to possible effects in higher animals on the fertility of either sex, the gestation or postnatal care, and development of the embryo. No direct causal relationships have been shown between any of the commonly used preservatives and reproductive toxicity. However, indirect reproductive toxicity has been suggested, for example, benzoic acid as a preservative, increases blood salicylate levels, and may cause a significant increase in the teratogenic potential of aspirin. Organic mercury has been shown to be responsible for acongenital form of

Minimata disease, but not at levels likely to be taken into the body through mercury-containing preservatives.

Carcinogenicity and Mutagenicity

Several preservatives have been scrutinized for being possible carcinogens, and their use regulated (e.g., chloroform and formaldehyde). Another possibility is that the preservative interacts with amines or amides to form carcinogenic nitrosamines. Published data on the mutagenicity of preservatives is limited as tests such as the Ames test can only be carried out at concentrations far less than those employed.

Preservative Combinations

The ideal properties expected from preservatives are not met by any of the current, established agents. All chemical agents have their limitations in terms of their antimicrobial activity, resistance to organic matter, stability, incompatibility, irritancy, or toxicity. It is also unlikely that new preservatives will be developed because of high costs of putting a new compound through toxicity tests and the likelihood of a new compound being impossible to cover by patent. More likely preservative combinations will be employed in the future, possibly to give synergistic combinations (e.g., hydrogen peroxide and peroxygen compounds, chlorhexidine and cetrimide). In this manner, preservative combinations may be used to extend the range and spectrum of preservation; for example, by combining a series of alkyl esters of 4-hydroxybenzoic acid, water solubility is decreased to such an extent that both the aqueous and oil phase of an emulsion are protected.

Table 7.5. Examples of interactions between preservatives

	Preservative B		
Preservative A	*Synergistic*	*Additive*	*Antagonistic*
Benzalkonium	Phenylmercurics	3-Cresol	Hexachlorophane
Benzoic acid	Dehydroacetic acid	Parabens	Boric acids
Bronopol	Benzalkonium	Sorbic acid	—
Chlorbutanol	Phenylethanol	—	EDTA, EGTA
Chlorocresol	Phenylethanol	—	—
Phenylmercurics	3-Cresol	—	—
Parabens	Germall	Benzoic acid	Boric acid
Sorbic acid	Dehydroacetic acid	Benzoic Acid	Parabens

When acting simultaneously on a microbial population, combinations of preservatives may cause an increased, decreased, or unchanged antimicrobial response when compared with their summed individual

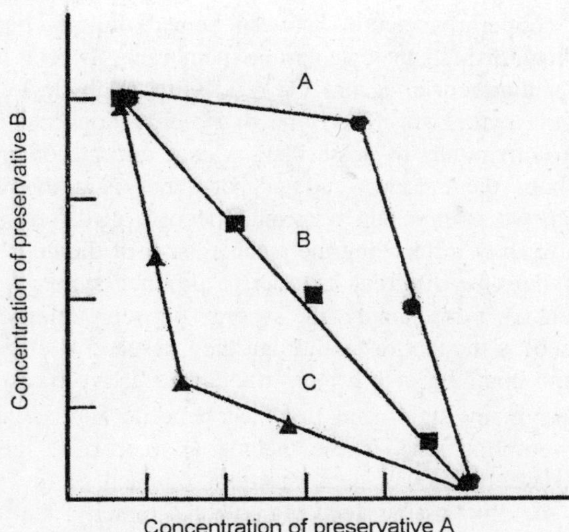

Fig. 7.4. Isobologram showing three possible outcomes of interaction between preservatives A and B, namely (A) antagonism; (B) additivity; and (C) synergy.

effects. A minimum requirement of any combination should be that it achieves at least the same level of protection overall as the individual components do. Ill-considered preservative combinations may lead to inclusion of irrelevant agents. In general, agents from the same chemical group or those having the same mode of action are likely to produce merely additive effects whereas those exhibiting different mechanisms or sites of action may serve either to reinforce (synergize) or reduce (antagonize) their individual activities. Such effects may be evaluated by preparing mixtures of the two preservatives being investigated and determining their growth inhibitory power by an MIC determination.

In general, synergistic effects can occur through three different mechanisms. The first mechanism is by inhibition of inactivation. Here, one compound increases the effective concentration of the other by inhibiting the microbial system responsible for its inactivation. To date, however, there are no reported observations of this mechanism with preservatives. The second synergistic mechanism is very common amongst preservatives and occurs when one compound increases the accessibility of targets to another. This usually occurs when one agent exerts its action by increasing the permeability of the cell wall or cell membrane. Besides damaging the cell in its own right, this can permit access (or increased access) of the second compound to targets that were previously concealed. An example of this type of synergy is seen in the cooperative action between benzalkonium chloride and thiomersal which disrupts the cytoplasmic membrane, thereby facilitating access of organomercurial agents that react with sulphydryl-containing enzymes in the cytoplasm. This type of synergy does not necessary require both compounds to possess significant antimicrobial activity. Finally, perhaps the broadest category of synergistic interactions is that in which two compounds act simultaneously at different targets, thereby significantly influencing the biochemistry of the cell. Although these targets must be different in order to produce synergy, they can be closely related. For example, the synergy between chlorocresol and 2-phenylethanol is thought to be through their respective effects on the generation and coupling of a proton gradient to active transport.

There are many significant practical benefits and advantages of preservative combinations. These include an increased spectrum of activity; the need for lower concentrations of each of the individual components, thereby possibly reducing potential toxicity; the prevention of resistance to individual preservatives; possible enhancement of antimicrobial activity beyond that expected by simple addition; an extended time course of preservation achieved by combining a labile, markedly biocidal preservative with a stable longer- acting ingredient. Furthermore, careful selection of individual agents for combination, based on their physicochemical properties, may serve to overcome microbiological problems created by the physical limitations of individual preservatives. In this respect, factors affecting preservative efficacy must be carefully considered.

8

PHARMACOKINETICS

Patient compliance may be improved when drug administration is tied to meal timing and intake. However, depending on the drug, the composition and the size of the meal, and the time of food intake relative to dose administration, the pharmacokinetic properties of the drug may be altered. These perturbations in the pharmacokinetics of a drug as mediated by food may cause a wide range of effects including increased drug absorption, prolonged time to peak concentrations, increased variability in drug exposure, decreased drug absorption, or inhibition of metabolism and elimination processes. Many times, these changes may adversely influence the safety and efficacy of the drug substance or product and they are highly undesirable. For new chemical entities that have been approved for marketing, the label alerts the patient and the physician by including a statement under Dosage and Administration on the extent of alterations in peak concentrations (C_{max}) and area under the plasma concentration vs. time curve (AUC), potential clinical significance and directions on time of dose administration relative to time of meal intake. The Food and Drug Administration (FDA) has issued a draft biopharmaceutics guidance for industry entitled "Food-Effect Bioavailability and Fed Bioequivalence Studies: Study Design, Data Analysis, and Labeling" in October 2001. However, a final guidance has not yet been issued to date.

A number of intrinsic and extrinsic factors influence the pharmacokinetics of a drug in both the fasted and the fed states. These include anatomical, pathological, and physiological factors as well as the physical and chemical properties of the drug, formulation, or excipients. Because of the inherent complexities in perturbations that food may cause in the pharmacokinetics, safety, and efficacy of a drug, food-effect assessments in early drug development are typically performed proactively by pharmaceutical companies to facilitate design of pivotal clinical efficacy studies and, eventually, the drug product label. This article will highlight the factors influencing the pharmacokinetics of the drug in the fasting and the fed conditions as well as the importance of performing food-effect studies in early drug development. This article will also provide a pharmaceutical industry perspective on the models that are presently available to predict the presence and the extent of food-mediated alterations in absorption, metabolism, and clearance as they relate to a drug development program. Although food intake can affect the pharmacokinetics of a drug for a number of routes of administration, the emphasis of this article will be on drugs administered via the oral route of administration.

Effects of Food on the Pharmacokinetics of a Drug

Food has a myriad of effects on the pharmacokinetic behavior of a drug substance regardless of therapeutic class. Key variables include the meal itself, the composition of the meal, the drug, the formulation,

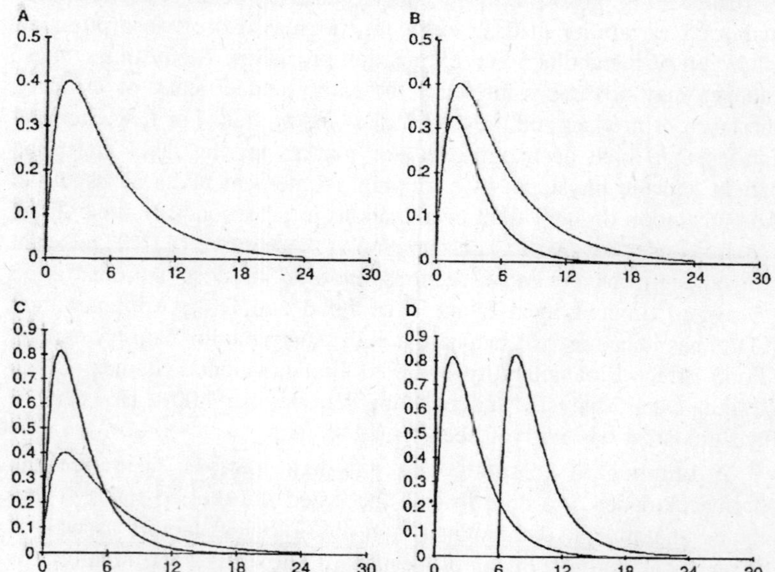

Fig. 8.1. (A) Representative plasma concentration vs. time profile of a drug A after oral dosing in the fasted state. (B) Food decreases exposure (C_{max} and AUC) of drug. (C) Food increases exposure (C_{max}) of a drug A. (D) Food delays exposure but does not decrease or increase exposure (C_{max} or AUC) of drug A.

the concomitant medications, the time of day, and the disease pathology. Such differences are often as a result of altered physiological influences mediated by food. Examples of these physiological effects include the degree to which the gastric emptying rate is reduced, the variations in the degree to which digestive enzymes are secreted, the dynamics of an altered pH and the consequences for ionized drugs in the gastrointestinal (GI) tract, the biliary excretion, and the stimulation of hepatic and/or splanchnic blood flow. Other important influences include physical interaction between the elements of food and the drug itself, which can interfere with absorption of a drug or merely act as a barrier to absorption from gut mucosa. The discussion below is primarily centered on how this knowledge can be integrated into rational drug development. Examples are drawn from case studies of drugs in drug development, the issues of identifying food effects early in the development, timing, design, and conduct of such studies as well as the issues and opportunities in this specific area of drug development.

Table 8.1. Partial list of drugs whose pharmacokinetics are influenced by intake of food

- Abacavir
- Adinazolam
- 5-Aminosalicylic acid
- Atorvastatin
- Avitriptan
- Bromazepam
- Bumetanide
- Celecoxib
- CGP 43371
- Clodronate
- Cyclosporin
- Danazol
- Didanosine
- Erythromycin
- Fexofenadine
- Furosemide
- Ganciclovir
- Halofantrine
- Inidnavir
- Itraconazole
- Levofloxacin
- Methotrexate
- Nifedipine
- Pravastatin
- Rifabutin
- Stavudine
- Tacrine
- UFT/leucovorin
- Venlafaxine
- Zolmitriptan

Human Food-effect Studies and Available Pharmacokinetic Methodologies

The variation in the size and intake of food, the type of food, ethnicity, the caloric value, the fat content, the time gap between meals, and the intergastric volume can influence the physiological environment wherein the drug may find itself upon administration. All of these factors may result in alterations in the bioavailability of a given drug such as stimulation of biliary secretion and splanchnic blood flow, changes in GI pH, or delay in gastric emptying. In order to capture these perturbations in bioavailability, the FDA recommends a high-fat and high-calorie meal for testing the effect of a meal on drug pharmacokinetics. Historically, this meal has been considered a "worst case scenario" for a drug–food interaction study. For new chemical entities submitted for approval in the United States, well-controlled pharmacokinetic studies are performed using this "*standard high-fat meal*" as the fed condition. One may argue that the above definition of standard high-fat meal is probably not representative of a standard breakfast in the United States or other parts of the world. Exceptions include vegetarian or vegan meals, low-fat/low-calorie meals for diabetics as recommended by the American Diabetic Association (ADA), and meals that are specific for each ethnic group. All of these differences compounded by genetic predisposition to pharmacokinetics may profoundly impact the pharmacokinetics of a drug and its prescribing habit. Nonetheless, keeping the type (or lack) of meal as a constant, one can eludicate the effect of other variables (formulation, time of meal intake relative to dose, etc.) on pharmacokinetics keeping in mind that the meal should be reflective of one that is used (recommended) during patient therapy.

There are two types of food-effect studies that are often performed in humans. These studies are designed to provide dosing recommendation with meals or to provide a mechanistic investigation on the effect of food on the pharmacokinetics of drugs. Examples of the former include studies done in conformity with the FDA guidance on food-effect studies or called "*standard assessments*." Examples of the latter include exploratory mechanistic studies to investigate gastric emptying using tools, such as gamma-scintiography, or modulating gastric pH using ranitidine coadministration, with the aim to provide/elucidate underlying reasons for observed food effects.

Standard Food–effect Assessments—regulatory and Product-labeling Considerations

The draft FDA guidance entitled "Food-Effect Bioavailability and Fed Bioequivalence Studies: Study Design, Data Analysis, and Labeling" provides sponsors with information on the design of appropriate studies to assess impact of food-mediated alterations in pharmacokinetics and to provide suitable dosing guidelines on drug product label. The guidance provides useful strategies for biowaivers if the drug in question belongs to a particular Biopharmaceutics Classification System (BCS) category. For example, for highly soluble and highly permeable drugs (i.e., BCS Class I), food effects should be inconsequential (between test and reference products) due to the inherent properties of the drug. The premise is that the compound is minimally influenced by food as a result of pH- or site-independent absorption and lack of dissolution differences.

A suitable study design for definitively assessing the influence of food on the pharmacokinetics of a drug or drug product includes performing a single-dose, randomized, balanced, two-period, two-treatment (fed vs. fasted) crossover study, with an adequate washout period between treatments. The number of subjects in the study should provide adequate statistical power for capturing small differences in exposure variables [peak concentrations ($C_{max)}$, T_{max}, and area under the plasma concentration vs. time curve (AUC)]. The guidance also asks for using the highest strength of the formulation to be evaluated in such studies. For the fasting treatment, at least a 10-hr fast prior to dosing is recommended: "Following an overnight fast of at least 10 hours, subject should take the drug product. No food should be allowed for at least 4 hours post-dose. Water can be allowed at libitum after 2 hours. Scheduled standardized meals should be served throughout the remaining study period." The issue is not the "number" of hours of predose fast time, but the stomach volume. If it contains <50 mL (fasting state depends on the volume present in the upper region of the stomach), the stomach is in an interdigestive state. It is generally recommended to have a postdose fast time of about 4 h. For the fed treatment, the regulatory guidance indicates a standard high-fat meal composed of two eggs fried in butter, two strips of bacon, two slices of toast with butter, 4 oz of hash brown potatoes, and 8 oz of whole milk. The caloric breakdown of this meal translates into approximately 150 cal from protein, 250 cal from carbohydrate, and 600 cal from fat. Standard pharmacokinetic calculations are performed using non-compartmental methods, e.g., $C_{max,}$ AUC, $T_{max,}$ etc. The guidance provides directions on how the data are applied to claim label attributes; for example, the degree to which 90% confidence intervals for the ratio of geometric means between fasted and fed legs are contained relative to the bounds of 80–125% for AUC or C_{max}.

One issue that routinely arises is how to provide the FDA-recommended high-fat meal at non-U.S. investigative sites, where this meal is not readily accessible. Simple substitution of locally available food (e.g., sausage for bacon) that maintains the caloric breakdown of the FDA-recommended high-fat meal has been an acceptable approach.

A second type of study involves applying food as a covariate in population pharmacokinetic models. This is an emerging area of interest that is gaining regulatory concurrence. The advantages of using multilevel population pharmacokinetic models include the determination of the extent of both interindividual and intraindividual variability, the use of a limited sparse sampling approach, and the use of multiple covariates within a large Phase III trial setting. This strategy allows the simultaneous analysis of relevant covariate relationships (negative vs. positive correlation, proportional correlation, etc.) influencing the key pharmacokinetic parameters as a function of body surface area, age, weight, sex, hematological or biochemical indices, creatinine clearance, and, more importantly, any delays in drug absorption as mediated by food intake. Such analyses can be performed using tools, such as NONMEM, with standard conditional estimation methods (first order, first-order conditional estimate, or first-order conditional estimate with interaction, etc.). This allows for estimation of typical values for key pharmacokinetic parameters, *V/F*, *CL/F*, absorption rate constant, or lag time. Shepard et al. used a sequential approach with an indicator variable to assess the effect of food on the parameter of interest. Their model included $K_a = \theta_1(1 - Q) + \theta_2 Q$, where Q was an indicator variable with a value of 0 (fasted) or 1 (fed), and θ_1 and θ_2 were the parameters of interest, with $K_a = \theta_1$ for fasted treatment and $K_a = \theta_2$ for fed treatment.

A number of recent examples in the literature describe the powerful use of population pharmacokinetic strategies to maximize information pertaining to food–effect behavior. These include a combination of non-compartmental and population pharmacokinetic analyses to answer a scientific question. For example, Zhou et al. used non-compartmental analysis to determine the extent of decrease in bioavailability of a drug mediated by food and then showed, using deconvolution and subsequent population pharmacokinetic analysis to assess the impact of meal timing on absorption, that the reduction in bioavailability under fed conditions was related to a reduction in the extent of absorption and not to a reduction in the rate of absorption. The use of the population pharmacokinetic approach in a multicenter trial in patients to assess food effect using sparse sampling and several categorical and continuous covariates is a more common application of this strategy.

However, it is not clear that a population pharmacokinetic approach definitively addresses a food interaction without a traditional Phase I study.

Scintigraphic Assessments

The technique of gamma scintigraphy has been used to evaluate biopharmaceutic properties of drug substances administered orally in the fed and the fasted states to evaluate food effects. Specifically, the procedure allows for studying GI transit of radiolabeled drug formulations in the fed state. With this method, one can understand the physiological processes by which the GI tract manages food and how the stomach empties meals of varying chemical composition. Such type of studies may also reveal the potential sites of absorption of an investigational product throughout the GI tract and how this is impacted in the fed state. This information is particularly useful when designing modified-release formulations. Technitium (^{99m}Tc) is a widely used nuclide in radiopharmaceuticals because of its optimal energy, short half-life, and small radiation dose needed.

Models for Assessing Impact of Food on the Pharmacokinetics of a Drug Substance

In Vitro Models

Biorelevant dissolution models have been described in the literature as a means to predict food-effect behavior on the absorption of drugs. Reppas et al. have suggested several dissolution media that simulate fasted and fed conditions in the human body. They used milk (3.5% fat) and United States Pharmacopeia (USP)-simulated gastric fluid with and without pepsin as dissolution media representing the fed and the fasted states in the stomach, respectively. Dressman and Reppas have showed that such biorelevant media have distinct advantages over compendial media in the prediction of food effects for poorly water-soluble drugs. An example of how dissolution tests may be used to predict the in vivo effect of food on drug absorption for a BCS class II (highly permeable, poorly water-soluble drug) drug, danazol. The dissolution profile of danazol in different media, namely, water, simulated intestinal fluid (SIF), FaSSIF, and FeSSIF. FaSSIF represents the fasted state in small intestine and FeSSIF represents fed state in small intestine. From these dissolution tests, it is clear that there is enhanced dissolution in the simulated fed state relative to the simulated fasted condition. Not surprisingly, the extent of this in vitro increase is similar to the extent of increase in danazol exposure in human

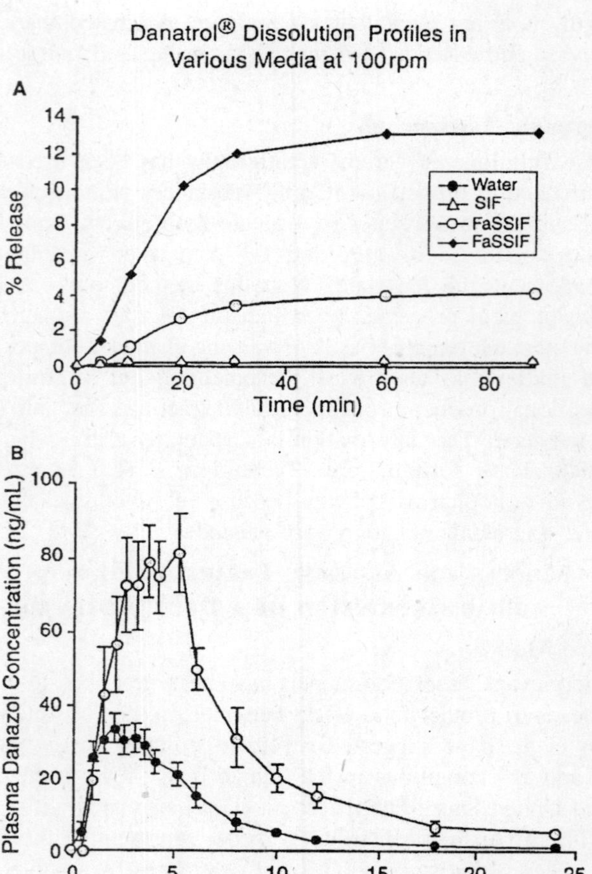

Fig. 8.2. (A) In vitro dissolution profile for Danazol. (B) In vivo plasma concentration vs. time profile for Danazol.

subjects observed in the fed state. The mechanism of food effect is related to increased solubilization mediated by bile salts for this steroid drug. This strategy appears to be predictive of drugs that belong to the BCS Class II category and may not be generally applicable to drugs belonging to other categories.

In Vivo Models

A number of animal models have been used to describe the fasting and the fed pharmacokinetics of drugs with the aim of predicting human performance. These include mice, rats, rabbits, guinea pigs, dogs, pigs, monkeys, to name a common few. The discussion below will address the relative merits and demerits of rodents, dogs, and pigs as predictive models for humans. It should be kept in mind that animal models may be helpful in assessing large and overt food interactions in oral bioavailability. However, relatively small differences that may be ultimately important in human studies and therapeutics will probably not be detected.

Rodents

Rodents (mice, rats, guinea pigs) are generally not preferred as models for assessing food effects for various reasons, namely, differences in GI anatomy and physiology, difficulties in administering "human" food, and lack of implementing a crossover design assessment. Other potential caveats include differences in the pharmacokinetic and the metabolism behavior in rodents as compared to humans and non-human primates that may negate assessment in this species. Metabolic profiling of the investigational product in a panel of animal models will reveal the suitability of a particular species as a predictor of human pharmacokinetic behavior. Furthermore, the extent of serial sampling is also constrained in smaller rodents such as mice. However, rodents are one of the two species commonly and typically used as a toxicology species in drug development; it is of importance to assess the impact of food not necessarily from the point of view of predicting human food effect but more so in ensuring adequate exposure multiples are obtained for toxicokinetic purposes. Few authors have explored rat as a model for predicting food-effect behavior.

Dogs

Pentagastrin-pretreated dogs have been shown to be useful in screening biopharmaceutic properties of new formulations. Coupled with the features that a crossover study design can be easily applied and human food easily administered to these animals, the dog remains one of the commonly used models in assessing the pharmacokinetics of drugs. The disadvantages of using a dog model include dissimilar GI physiology and, hence, differences in pharmacokinetic behavior (for the pentagastrin-untreated dogs), need for modulation of GI physiology using pentagastrin and other physiological agents, and dissimilar drug metabolism/transport between dogs and humans for some compounds. Additionally, dogs are relatively more sensitive to physiological effects of certain pharmaceutical excipients commonly employed in solubilization of poorly water-soluble drugs (e.g., Polysorbate 80 and resultant orthostatic hypotension).

Although gastric emptying time is relatively similar between dogs and humans in the fasted state, the intestinal transit time is almost twice as long in humans. These apparent differences can be offset by administering pentagastrin pretreatment which allows the gastric acidity, the emptying time, and the small intestinal transit time to closely approximate human conditions, i.e., approximately pH 2, 0.7 and 4 hr, respectively. Another approach to synchronize the gastric environment in dogs to closely mimic humans includes a combined treatment of intramuscular pentagastrin with intravenous atropine sulfate.

Pentagastrin is an analog of gastrin that reproducibly stimulates gastric acid secretion and is usually administered intramuscularly prior to biopharmaceutic experimentation in doses of approximately 6–10 μg/kg. Conflicting results in the literature of drugs studied in these pretreated dogs have negated the extensive use of this model. For example, pentagastrin-pretreated dogs were more predictive of the human bioavailability of didanosine than the untreated dogs. In contrast, data from Kanerva et al. suggest that the pentagastrin-pretreated dog was unable to reliably predict the absorption of deramciclane, an acid-labile drug.

A typical design of a food-effect study in dogs is as follows. Beagle dogs (usually n = 6–8) are administered a pharmacological dose of a drug formulation (hard gelatin capsule or tablet) in a randomized or non-randomized crossover design. Treatment legs may include dosing after an overnight fast (fasted state) and dosing with a low-fat meal, a high-fat meal, or other specialized meals (fed state). The composition of meals that have been administered to dogs can be highly variable. For example, Paulson et al. administered low-, medium, and high-fat diets. The low-fat diet was composed of one slice of toasted white bread spread with 0.5 oz of jelly, 8 oz of skim milk, and 6 oz of orange juice. The medium-fat diet was composed of one slice of toasted white bread with 0.5 oz each of peanut butter and jelly, 1 oz of dry cereal (cornflakes), 8 oz of skim milk, 6 oz of orange juice, and one banana. The high-fat diet was composed of two slices of toasted white bread spread with 1.2 oz of butter, two eggs fried in butter, two slices of cooked bacon, 2 oz of hash brown potatoes fried in butter, and 8 oz of whole milk. These diets were homogenized then administered either in a bowl or with a syringe to the dogs. A suitable washout period separated the treatment legs. Blood samples are collected at predetermined intervals predrug and postdrug administration and pharmacokinetic behavior assessed with and without food.

Pigs and minipigs

The digestive tract of a minipig is anatomically and physiologically similar to that of the human adult digestive tract. In fact, the weanling minipig model has a GI tract that is closely similar to children between the age of 2 and 5 years. Specifically, the similarities exist for the gastric villi, the regulation and the composition of gastric secretions, the transit times as well as in the digestive and the absorptive behaviors. Regardless of these advantages, the use of the minipig in pharmacokinetic assessments has been relatively uncommon. One example of an application to the assessment of food effects in miniswine is the study on the effect of various fat-content diets on the bioavailability of theophylline following a 400-mg single-dose Theo-24. The observed similarities in the pharmacokinetics of theophylline in pigs and humans led Koritz et al. to suggest swine as a predictive model for the assessment of theophylline bioequivalence in humans.

Food-effect Assessment in Drug Development

Timing of Food-Effect Studies

There are generally two types of clinical pharmacokinetic/pharmacodynamic (PK/PD) studies that may be performed during drug development. One is a "pilot" study to obtain preliminary exploratory assessment used for facilitating internal decision making for the progression of a compound and the other is a more formal statistically powered "pivotal" study to provide dosing recommendation for purposes of the drug product label. The key difference is that the pilot study is often powered for large differences and the pivotal study is powered for small differences for a priori defined endpoint. The pilot study, which usually uses a developmental formulation and a guess of the therapeutic dose, can also provide valuable information for developing a well-designed pivotal study. Label-driven food-effect assessments are performed in later stages of drug development, particularly when the "market" formulation is available and the therapeutic dose is known from Phase III efficacy studies. Appropriate recommendations for performing such studies are available in the Guidance for Industry issued by the FDA in the United States.

The exploratory assessment of food on the pharmacokinetics of new chemical entities in the drug development chain of events varies largely within companies. This is also driven by marketing considerations, competitive advantage, flexibility for dosing options, and the therapeutic index of the drug if known. Several authors have suggested inclusion of an exploratory assessment early in the drug

Table 8.2. Typical clinical pharmacology and PK/PD studies in early drug development

I. Clinical pharmacology studies
Single-ascending-dose safety, PK, and PD
Multiple-ascending-dose safety, PK, and PD
Proof of mechanism study
Proof of principle study
II. Clinical PK/PD studies
Age and gender PK
Drug–drug interaction
Renal and hepatic impairment PK
BA/BE food and fasted PK
Dose proportionality and diurnal variation
Single- and multiple-dose PK/PD
14C ADME study
Tissue distribution (PET, gamma scintiography, etc.) PK

development program because appropriate formulation development may be initiated for drugs exhibiting undesirable food effects. A caveat for this approach is the large number of food-effect studies that may need to be performed if multiple changes are made for the formulation and/or the excipients within them. If multiple-formulation switches occur in a program, it may be worthwhile to delay the food-effect assessment to discount the formulation as a variable.

A convenient method to examine food effects is the inclusion of a food-effect assessment as part of the first-in-man study once the *maximum tolerated dose* (MTD) is identified. Depending on the drug bioavailability characteristics in animals, physicochemical properties of the drug and preliminary data on whether the drug exhibits dissolution rate-limited absorption behavior, a dose that is 1/2 to 1/4 of the MTD, may initially be used for the preliminary food-effect assessment. This method potentially avoids undesirably high exposures of the drug when adequate safety data are not available. Dose selection will also depend on the predicted therapeutically relevant dose range, the linearity in pharmacokinetic behavior, and the sensitivity of the bioanalytical assay. Typically, the FIM study design includes non-randomized, non-crossover, single-ascending doses in eight subjects (six active, two placebo), where doses are escalated until an MTD is reached and dosing typically performed after an overnight fast. For example, in the food-effect leg, 0.25–0.5 × MTD cohorts return for a second dose after a washout period wherein dose is administered with a standard high-fat meal. This allows within-subject comparison of exposure with and without food, but does not account for period effects. Hence, it is not unusual to obtain inconclusive results with this design. To lend rigor in the scientific assessment of food effect, it may be more desirable to have the food-effect assessment done as part of the *first-in-man* (FIM) study using a randomized crossover design in eight active subjects.

In general, however, it is recommended to perform specific well-designed food-effect studies with adequate power for large differences to capture meaningful deviations in exposure.

Single vs. Multiple Dosing—Dose Selection

The selection of the appropriate dose in assessing the effect of food on the pharmacokinetics of a drug is an important consideration. Dose selection may be based on several criteria, namely, therapeutically relevant concentrations (if known), safety margin, linear pharmacokinetics, etc. It is generally required to evaluate a single dose of a drug for such an assessment. However, depending on non-linear pharmacokinetic behavior and the degree of accumulation (or induction) upon multiple dosing, a multiple dose may be a better option. A multiple-dose study may also be more meaningful if the primary or secondary objective is to evaluate the effect of food on the safety or the pharmacodynamics of a drug, where single-dose administration may not be sufficient to induce meaningful perturbations in pharmacodynamics. An example of this approach is in the multiple-dose administration of the "statin" group of antihyperlipidemic agents for assessing the effect of food on the lipid-lowering efficacy [i.e., LDL-cholesterol (LDL-C) levels] given the slow turnover rate of LDL-C.

Dose selection based on pharmacokinetic profile

The oral absorption of certain highly permeable, poorly water-soluble drugs with low oral bioavailability in the fasted state may be profoundly influenced by the intake of fatty meal. Dose selection for such drugs can be challenging and important. For example, acitretin is a retinoid indicated for the treatment of psoriasis. Depending on the dose of acitretin used, food appears to increase the absorption in a varying manner. In the fasted state, both the C_{max} and the AUC reach a plateau at 50-mg dose with minor, if any, increases in exposure at doses higher than 50mg. In contrast, the drug exhibits almost dose-proportional increases in C_{max} and AUC over the entire tested dose

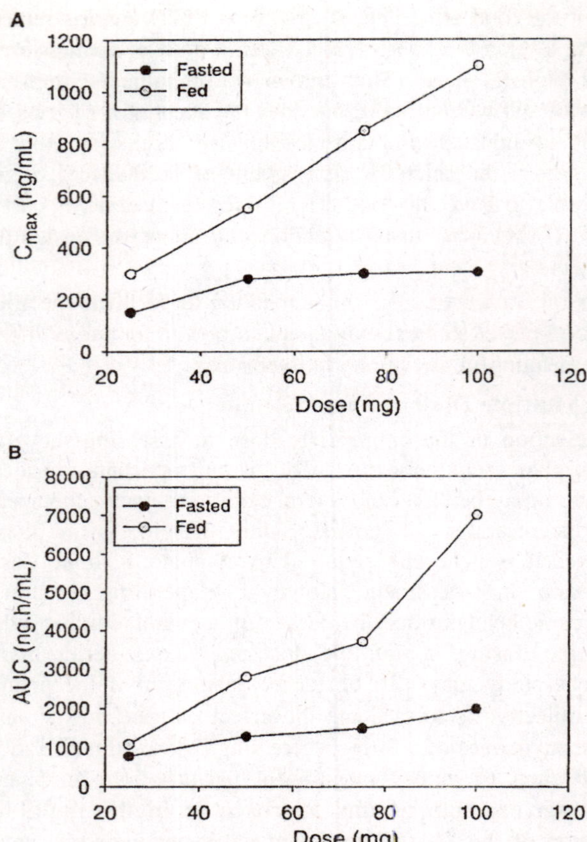

Fig. 8.3. (A) Relationship between dose and C_{max} for acitretin. (B) Relationship between dose and AUC for acitretin.

range up to 100 mg, when the drug is administered with a high-fat meal. When acitretin is given with food, there is also a decrease in inter subject variability in exposure parameters. These results have important implications for investigational agents exhibiting dissolution rate-limited absorption during drug development as this presents an opportunity for key formulation development to increase bioavailability if safety is not an issue and if formulation options are available. This example also illustrates the need for an early assessment of a food interaction study. In acitretin, the patients in the Phase III pivotal studies were instructed to take their daily acitretin dose with the main meal of the day based on this information.

Dose selection based on PK/PD relationship

A second more germane consideration is dose selection based on PK/PD relationships. Depending on the reference point on a PK/PD chart, food may have little or profound effect on either PK, PD, or both. For example, drug A shows profound reductions in C_{max} and AUC, but blood pressure reduction is relatively unaffected by food; it is conceivable that the drug PD is at the maximal effect where large differences in PK leads to minor changes in PD. On the other hand, if the dose of drug B was selected such that its PD is at 50% of maximal effect, then there is much larger PD differences for observed differences in drug exposure.

An example of the former case (drug A behavior) is illustrated in the effect of food on the PK and the PD of pravastatin, a representative statin in the treatment of hyperlipidemia. Although a low-fat, low-cholesterol meal altered the pharmacokinetics of 20-mg pravastatin significantly, no adverse impact on pharmacodynamics (lipid-lowering efficacy) was observed. Concomitant intake of meals reduced pravastatin C_{max} by 49% and AUC by 31%, the change in LDL-C-lowering efficacy was minimal (–37% with meal vs. –36% 1-hr premeal). The authors recommend administering pravastatin without regard to meals. It is worth noting here that the LDL-C-lowering efficacy of pravastatin plateaus at ca. –32% (20 mg) to ca. –37% (80 mg) changes from baseline; at –37%, the PD is already at the flat portion of the sigmoidal PK–PD curve. Consequently, even with a change in PK approximating

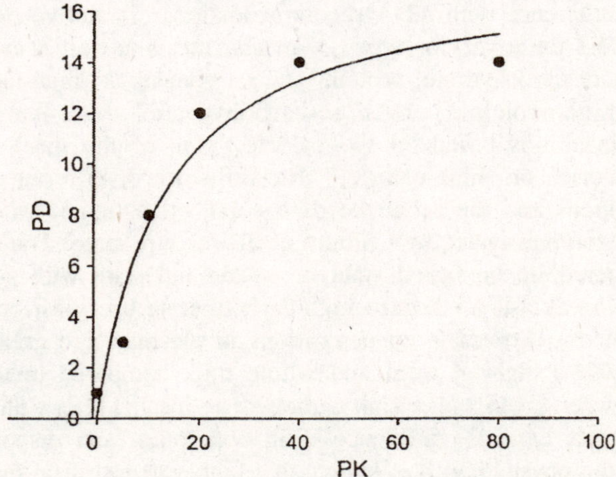

Fig. 8.4. Representative illustration of PK/PD relationship for a hypothetical drug.

50%, there is minimal impact on PD. It is unlikely that the PD would be influenced significantly by food for compounds whose PD is at this flat portion of the PK–PD curve than would be the case if the dose selected represented a PD of 0.5 times maximal effect, where there is a much larger variability for PD differences with variability in PK.

Type of Meals

Another key consideration when performing food- effect studies is the type of meal utilized for assessing effect of food on the pharmacokinetics of a drug. Because of differing ethnicities, geographical regions, and variations in the eating behavior of human beings across the world, it is conceivable that the "standard" breakfast, lunch, and dinner will vary extensively. Then it becomes increasingly difficult to assess the true significance of food and its impact on PK/PD and to develop dosing recommendations that are systematically applied to a drug that is prescribed globally.

In order to provide a systematic basis for capturing food-mediated alterations in the pharmacokinetics of drugs, the FDA draft guidance recommends a standard high-fat meal as being composed of "2 eggs fried in butter, 2 strips of bacon, 2 slices of toast with butter, 4 ounces of hash brown potatoes, 8 ounces of whole milk." Differences in pharmacokinetic behavior may be expected depending on the type of breakfast utilized; for example, an English breakfast is different in composition and caloric value to an American breakfast or a special diet in adherence with ADA recommendations. To derive meaningful data across these varying types of meals, meals are often categorized in terms of carbohydrate, protein, and fat content as "high-fat," "low-fat," "high-protein," or "high-carbohydrate" diet. An alternate categorization is based on the caloric value of the meal itself, as "low calorie" or "high calorie." The influence of differing breakfast compositions and the influence of a solid vs. a liquid (whole milk) meal were investigated by Colburn et al. for etretinate. The drug was administered in the fasted state or in the fed state with a standard high-fat breakfast, a standard high-carbohydrate breakfast, and whole milk (16 oz). Etretinate concentrations in plasma were greater when given with a high-fat meal and whole milk compared to the fasted state or when given with a high-carbohydrate meal. Meals with differing fat contents can also influence drug absorption and disposition by influencing physiology. It is known that higher levels of fat may inhibit gastric emptying. Moreover, the GI transit times are different for solid and liquid meals and this may impact the pharmacokinetics of a drug. The lack of consistency in the scientific literature on the type of meal used in the investigation prohibits data comparisons across populations unless such differing compositions are used in a single study.

Table 8.3. Partial list of ingredients of meals that influence drug absorption, metabolism, and elimination

- High-fat-containing products
- Juices
  - Orange juice
  - Grapefruit juice
  - Apple juice
- Red wine
- Black pepper
- Tea
- Garlic
- Broccoli

Of more importance in relation to the influence of food on absorption is the rate of blood flow in the capillary system. Food may cause an increase in splanchnic blood flow leading to alterations in drug absorption, depending on the ingredients present in the food that is ingested. Closely tied into this principle is the extent to which a given drug is extracted by first-pass metabolism when traversing through the liver via the portal circulation. Both the type and the size of meal can influence splanchnic blood flow. Using a non-invasive technique for the assessment of hepatic haemodynamics using radiocolloid imaging, Walmsley et al. assessed the influence of meal on hepatic, mesenteric, and splenic arterial effective blood flows. They showed that both the total effective liver blood flow and the mesenteric arterial effective flow increased by 28–69% after eating. It is known that long-chain fatty acids stimulate human intestinal blood flow.

It has become increasingly evident that several ingredients in food or components of a meal can influence drug metabolism and clearance via inhibition of cytochrome P450 isozymes or modulation of drug transporters such as p-glycoprotein. These ingredients include vegetables and fruit products and components such as St. John's wort, garlic, grapefruit juice, certain type of orange juices, apple juice, red, black, or white pepper. Red wine and green tea also have ingredients that can alter the metabolism/ clearance of a drug. In order to differentiate these influences from those arising from the altered GI physiology, the drug development program may need to include separate food and metabolism–transporter interaction studies to assess the overall impact on dosing recommendations.

Time of Meal Consumption Relative to Dose

The time of meal consumption relative to dose has important implications for compounds whose pharmacokinetics are severely impacted by food. The extent of food-mediated alterations in pharmacokinetics are influenced by the time interval between ingestion of food and intake of drug. Notably, as the time elapsed between meal ingestion and drug intake becomes longer, the less severe the effect of food on the pharmacokinetics of a drug would presumably be. Consider the following example.

If one were to fix the meal intake as a constant and to vary the time of intake of a meal, then differences in the rate and the extent of drug absorption may be observed. The treatments were: (A) 10-hr fast; (B) drug intake 1 hr before breakfast; (C) drug intake immediately following consumption of the breakfast; and (D) drug intake 1.5 hr after beginning consumption of the breakfast. As expected, tacrolimus absorption was the highest in the fasted state. The lowest impact on bioavailability was when taking the drug 1 hr prior to a meal; the highest impact was when tacrolimus was ingested immediately after a meal (treatments C and D).

Marathe et al. have investigated the effect of food on the pharmacokinetics of avitriptan in a systematic manner as part of the avitriptan drug development program. In the first study, they showed that concomitant ingestion of food significantly reduced avitriptan bioavailability. Specifically, mean C_{max} and AUC were reduced 70% and 45%, respectively, while T_{max} was delayed from 45 min to 2 hr when avitriptan was administered as a 50 mg capsule 5 min after a standard high-fat meal. Because avitriptan was developed as a treatment for migraine, this dramatic food effect has important implications for the developability of the drug for this indication. In the second study, Marathe et al. designed a meal-timing study where avitriptan was administered after an overnight fast and at 0.5, 1, 2, and 4 hr after intake of a standard high-fat meal. There was a meaningful relationship between avitriptan bioavailability as a function of time interval between dose and meal intake with the impact of food decreasing over time. Mean C_{max} decreased by 61%, 58%, 50%, and 35% after 0.5-, 1, 2-, and 4-hr postmeal, respectively; mean AUC decreased by 35%, 31%, 34%, and 19% after 0.5-, 1, 2-, and 4-hr postmeal, respectively. Median T_{max} values were 0.5 hr (fasted), 1.5 hr (0.5- and 1-hr postmeal), 1.25 hr (2-hr postmeal), and 0.75 hr (4-hr postmeal). In a later study by the same authors, scintigraphic analysis showed that avitriptan was rapidly absorbed from the upper small intestine after emptying from the stomach. Gastric emptying was slow and continuous in the fed state leading to extended absorption. The authors also attributed the high intrasubject variability in C_{max} and AUC to variability in gastric emptying under fasted and fed conditions.

A similar study was reported by Laitinen et al. on the effect of time of meal on the bioavailability of the bisphosphonate drug clodronate. Exposure (AUC) was the highest when clodronate was administered 2-hr premeal, followed by a 1-hr premeal (91% of 2-hr premeal AUC), a 0.5-hr premeal (69% of 2-hr pre-meal AUC), with meal (10% of 2-hr premeal AUC), and a 2-hr postmeal (34% of 2-hr premeal AUC). Although the 2-hr premeal does not present a "true" fasting state, there was a measurable difference between these short time intervals premeal and postmeal leading the investigators to conclude that clodronate be administered at least 0.5 hr prior to breakfast. The closeness of this prescribed time of drug intake to the meal ensures some balance of patient compliance with minimally acceptable compromise on the impact on drug absorption.

For a drug with a wide therapeutic margin, such variations in drug absorption relative to timing of a meal may be inconsequential in terms of clinical relevance; however, the same cannot be said for a drug with narrow therapeutic margin.

Type of Formulation

The type of formulation used, the excipients contained in the formulation as well as the particle size of the drug substance all may influence drug absorption and, thus, confound the observed food-effect behavior for a given drug. For example, a highly water-soluble drug may be formulated in a tablet that contains a poorly water-soluble polymer excipient that influences release of the drug from the matrix and subsequent dissolution properties of the tablet. Certain hydrophilic polymers, such as hydroxypropylmethyl-cellulose, have been implicated in dose dumping of modified- or controlled-release formulations in the presence of food. The use of varying excipients within formulations are often the cause of conflicting data for observed food effects. Whereas some excipients, such as mannitol, influence drug absorption by altering transit times, others, such as PEG-300, Cremophor EL, and Polysorbate 80, inhibit p-glycoprotein and other drug transporters.

In a drug development program, it is sometimes necessary to change the formulation depending on drug development timelines and availability of drug substances to meet the needs of clinical pharmacology and PK/PD studies in a timely manner. For example, the formulation may change from an aqueous suspension in the first-in-man study to a capsule formulation in proof-of-concept study to, eventually, a market formulation that would be a tablet formulation. Each time such formulation changes are made, a new food-effect study may need to be performed to assess the influence of the altered formulation on the observed food effect to design the Phase III study in an optimal manner. To minimize the number of food-effect studies performed due to formulation changes, it is essential to maintain consistent formulation strategy across the drug development phase.

Inclusion of excipients should occur based on informed input from a variety of functional areas including pharmaceutical sciences, clinical pharmacology, clinical PK/PD, and drug safety. A pivotal food-effect study needs to be performed using the same formulation used in Phase III pivotal efficacy trials or the market formulation so that appropriate dosing recommendations with regard to meals are provided.

Site of Drug Absorption

Absorption of drugs from the GI tract differs extensively depending on the intrinsic properties of the drug. For example, a drug may be absorbed from the upper GI tract or throughout the GI tract. This will have important implications for food effects as well as for other applications such as developing modified-release formulations or demonstrating bioequivalence of a test formulation vs. the reference formulation. For a drug that has an absorption window of the upper GI tract, a delay in T_{max} as mediated by food may lead to suboptimal absorption.

It is very clear that food can have significant pharmacokinetic and pharmacodynamic consequences for drug substances. Although not discussed in this article, food can also have deleterious effects on efficacy and safety of a product. A growing body of literature now points at many pharmaceutical excipients that are typically used in drug development cycle to influence drug disposition by modulating metabolism and efflux properties of drugs. Whether the information on food effects is derived for internal company decision making and developability considerations or for dosing recommendations with the final market formulation, understanding the myriad and often complex actions mediated by food is critical in drug development. A rationale selection of available choices within the context of a well-designed and adequately powered clinical study will provide important clues for the drug development scientist to: (i) enable well- designed late-stage clinical efficacy trials; (ii) optimize the formulations for early- and midstage developmental compounds; and (iii) provide prescribing options to the physician for optimal patient compliance on soon-to-be-marketed products.

9

Strategies for Molecular Farming

Plants have numerous advantages over traditional expression technologies for the large-scale production of recombinant proteins. The major benefits of whole plants include the comparatively low cost of large-scale production, the inherent scalability of agricultural systems and the convenience of existing infrastructure for harvesting, processing and distribution. The start-up and running costs for molecular farming in plants are significantly lower than those of cell-based production systems because there is no need for fermenters or the skilled operators to run them. It has been estimated that proteins can be produced in plants at 2–10% of the cost of microbial fermentation systems and at 0.1% of the cost of mammalian cell cultures or transgenic animals, as long as adequate yields can be achieved.

All plant expression platforms, including whole plants and cell/tissue culture systems, also have safety benefits over microbial and animal cells. This is because they lack the endotoxins produced by bacterial cells and they do not harbor human pathogens. Proteins produced in mammalian cells and transgenic animals must be screened for viruses, oncogenic DNA sequences and prions, which adds significantly to processing costs. Indeed, regardless of the production system, over 85% of the costs associated with recombinant protein production reflect extraction and down-stream processing steps rather than the production phase itself. For many proteins, however, these steps can be circumvented or eliminated if plants are used as production hosts. For example, recombinant subunit vaccines produced in plants can be administered orally as raw or partially processed fruits and vegetables, therapeutic proteins designed for topical application can be applied as crude extracts or pastes, and industrial enzymes used in food and feed processing can be expressed in the plant that needs to be processed. Even for those proteins that must be extracted and purified using conventional methods, plant systems can provide practical advantages to facilitate processing and make it more economical. This allows recombinant proteins to be concentrated in the oil bodies of oil-seed crops and extracted using a simple and economical method. Another example is the rhizosecretion of recombinant proteins from the roots of tobacco plants, which allows the protein to be collected continually from root exudates.

Given the multitude of benefits listed above, it is not surprising that many of our crop species have been investigated as potential hosts for molecular farming. More than 30 species of plants have now been transformed for the sole purpose of expressing and exploiting recombinant proteins, and it is becoming increasingly difficult to choose which expression host is the most suitable for particular products. Further complications are added by the diversity of expression systems available for each crop species. These may include transgenic plants, transplastomic plants, virus-infected plants, transiently transformed leaves, hairy roots and suspension cell cultures. There is also a wide choice of expression strategies, including leaf expression, seed expression, fruit expression, inducible expression, targeting to different subcellular compartments and secretion.

The choice of host species, expression system and expression strategy must be evaluated carefully on a case-by-case basis, depending on many inter-related factors. Strategic choices may be made for geographical reasons, such as the site intended for production and the local availability of labor, processing infrastructure, storage facilities, transport and distribution networks. The value of the recombinant protein is also important, since it would make economic sense to produce a high-value protein such as a recombinant antibody in an expensive expression system if other benefits were gained, while bulk products with a low market value would better be produced in a less expensive expression system. The widest choice is generally available where the recombinant protein needs to be purified to homogeneity, while there are greater constraints for proteins intended to be delivered in unprocessed or partially processed plant material. For example, proteins expressed for the purpose of oral vaccination in humans need to be expressed in the edible parts of food crops (e.g. tomato or banana fruits) while those intended for the oral vaccination of animals need to be expressed in fodder crops (e.g. alfalfa or clover leaves). Additional factors that may need to be taken into consideration include the degree of containment afforded by the crop and the extent to which the structure and homogeneity of N-linked glycans can be controlled. Different host species also vary in the expediency and convenience of transformation and regeneration, the availability of useful regulatory elements to control transgene expression, the extent to which endogenous compounds interfere with downstream processing, and in the absolute yields of recombinant proteins that can be achieved.

This chapter provides an overview of the different expression hosts, systems and strategies available in molecular farming, and discusses their advantages and disadvantages for the production of different types of recombinant proteins.

Host Species for Molecular Farming

Leafy Crops

The two major leafy crops used for the production of recombinant proteins are tobacco and alfalfa, both of which have high leaf biomass yields in part because they can be cropped several times every year. The main limitation of such crops is that the harvested leaves tend to have a restricted shelf life. The recombinant proteins exist in an aqueous environment and are therefore relatively unstable, which can reduce product yields. For proteins that must be extracted and purified, the leaves need to be dried or frozen for transport, or processed immediately after harvest at the production site. This adds considerably to the processing costs.

Tobacco (Nicotiana tabacum)

Cultivated tobacco has a long and successful history in molecular farming. The popularity of this species reflects its dual status as a model plant and a cultivated crop, providing many practical advantages for the large-scale production of recombinant proteins. As a model plant, tobacco benefits from well-established gene transfer and regeneration methodologies, and the availability of many robust expression cassettes for the control of transgene expression. The practical advantages of tobacco include its high biomass yield (up to 100 tonnes of leaf biomass per hectare each year), the wide range of available expression systems (transgenic plants, transplastomic plants, virus-infected plants, transient expression in leaves, transformed suspension cell cultures, hairy roots and shooty teratomas), and the fact that tobacco is neither a food nor a feed crop, thus reducing the likelihood of transgenic material contaminating the food or feed chains. The first recombinant human therapeutic protein to be produced in transgenic plants was expressed in tobacco leaves as was the first plant-derived recombinant antibody, the first plant-derived vaccine candidate, the first plant-derived industrial enzyme and the first plant-derived synthetic biopolymer. However, there are several drawbacks to molecular farming in tobacco leaves, particularly for pharmaceutical proteins. Many tobacco cultivars have high contents of nicotine and other alkaloids, which must be removed during the downstream processing steps. Even where low-alkaloid cultivars are used, processing is still necessary to remove these toxic components. It has also been shown that glycoproteins produced in tobacco are very heterogeneous in terms of their N-glycan structures, which could make tobacco unsuitable for the bulk production of certain pharmaceutical proteins.

Tobacco (Nicotiana benthamiana)

N. benthamiana is a non-cultivated tobacco species whose main advantage as a host plant for molecular farming is that it supports the systemic replication of many different viruses. This has two major applications. First, *N. benthamiana* is a suitable host species when the aim is to produce recombinant proteins using viral vectors, such as *tobacco mosaic virus* (TMV) and *potato virus X* (PVX). Second, transgenic *N. benthamiana* plants are often used to produce antibodies that are active against plant viruses, allowing the transgenic plant to be used as a viral assay system. While this is not strictly molecular farming (in that the aim is to confer disease resistance on the plant rather than to exploit the recombinant protein as a pharmaceutical or industrial product), the expression of antibodies in plants is of general interest because strategies to improve antibody expression levels or target them to specific compartments are also applicable to pharmaceutical proteins. *N. benthamiana* is occasionally used for molecular farming without the involvement of viruses, as in the production of a V_H domain recognizing the neuropeptide substance P.

Alfalfa (Medicago sativa)

The leaf biomass produced by alfalfa is somewhat lower than that of tobacco (12 tonnes per hectare per year), but it has several advantageous agronomic characteristics compared to tobacco including the fact that it is a perennial plant (vegetative growth can be maintained for many years), it can be clonally propagated by stem cutting, and it fixes its own nitrogen thus eliminating the need for fertilizer input. The technology for gene transfer and transgene control in alfalfa is not so well established as it is in tobacco, but due mainly to the efforts of researchers, much progress has been made in the development of constitutive and inducible expression cassettes for molecular farming, and alternative expression systems such as agroinfiltrated leaves and cell suspension cultures. While alfalfa lacks toxic alkaloids, the leaves do contain high levels of oxalic acid, which can in some cases interfere with downstream processing. The major advantage of alfalfa for the production of pharmaceutical proteins is that glycoproteins expressed in alfalfa leaves have homogeneous glycan structures. This, together with the ease of clonal propagation, provides enormous production benefits in terms of batch-to-batch reproducibility. Since alfalfa is a fodder crop, the other major application of this species in molecular farming is the delivery of vaccines to domestic animals, as has been demonstrated in the case of a foot-and-mouth-disease vaccine. Alfalfa has also been used for the production of three industrial enzymes: 1,4-β-D-endoglucanase and cellobiohydrolase, and phytase.

White clover (Trifolium repens)

Like alfalfa, the clover family (*Trifolium* spp.) consists of perennial legume plants that fix their own nitrogen. However, they suffer several limitations as general production crops for molecular farming, such as restricted perenniality, low protein content and the presence of high levels of condensed tannins which interfere with protein extraction. Despite these disadvantages, clovers are forage crops and are therefore useful for the production of vaccines against animal diseases. Thus far, white clover has been used to produce a *Mannheimia haemolytica* A1 leukotoxin fusion protein antigen to protect cattle against pneumonic pasteurellosis, otherwise known as *shipping fever*. The antigen remained stable for over one year in dried clover leaves that had been baked in the oven at 50°C immediately after harvest.

Lettuce (Lactuca sativa)

Lettuce has been used for the production of a hepatitis B surface antigen and was chosen essentially because it is an edible salad crop that can be used in human clinical trials. It does have a higher biomass yield than alfalfa (30 tonnes per hectare per year) but it has a much higher producer price and a very high water content (98%) which reduces protein yields and stability. Human volunteers, fed with transgenic lettuce plants expressing hepatitis B virus surface antigen, developed specific serum-IgG responses to the plant-derived vaccine.

The investigators who published this original report are now producing further pharmaceutical proteins in lettuce, including anthrax protective protein and antibodies against rabies and colorectal cancer. Other investigators are considering the use of lettuce for the production of further antigens, including most recently the SARS virus spike glycoprotein.

Spinach (Spinacia oleracea)

Spinach, like lettuce, has been used for the production of edible vaccines. In the first report, Yusibov and colleagues used *alflafla mosaic virus* (AlMV) to produce rabies fusion epitopes on the virus surface in infected spinach leaves. This resulted in the development of anti-rabies (as well as anti-AlMV) antibodies when the spinach leaves were fed to mice. Vaccines against HIV gp120 and Tat have been produced in spinach, and a construct of gp120 with the CD4 receptor is now being adapted for this plant. Spinach is also being used to make an anthrax vaccine.

Lupin (Lupinus spp.)

Lupin (*Lupinus luteus*) has been used to express the same hepatitis B surface antigen as produced in lettuce, and the transgenic leaves were used in pre-clinical trials in an attempt to promote an immune response in mice following oral administration. Narrow-leaf lupin (*Lupinus angustifolius*) has been used to express a gene encoding a plant allergen (sunflower seed albumin), in a successful attempt to suppress experimentally-induced asthma. This was the first study involving the production of a heterologous vaccine in plants to provide protection against allergic diseases.

Dry Seed Crops

The dry seed crops that have been used as host plants for molecular farming include the cereals maize, rice, wheat and barley, and the grain legumes soybean, pea, pigeon pea and peanut. Maize, rice, wheat, barley, soybean and pea have been investigated as general production platforms, while pigeon pea and peanut have been used solely for the expression of animal vaccine candidates. The major advantage of all seed crops is that recombinant proteins can be directed to accumulate specifically in the desiccated seeds. Therefore, although seed biomass yields are smaller than the leaf biomass yields of tobacco and alfalfa, this is offset by the increased stability of the proteins. Seeds are natural storage organs, with the optimal biochemical environment for the accumulation of large amounts of protein. In the best cases, recombinant proteins expressed in seeds have been shown to remain stable and active after storage at room temperature for over three years. The accumulation of proteins in the seed rather than vegetative organs also prevents any toxic effects on the host plant. Finally, the extraction of proteins from seeds is facilitated because the target protein is concentrated in a small volume, most cereal seeds lack the phenolic compounds that are often found in leaves and which interfere with processing, and the seed proteome is fairly simple, which reduces the likelihood that contaminating proteins will co-purify with the recombinant protein during down-stream processing.

Several factors need to be weighed up when choosing an appropriate dry seed expression host, including geographical considerations, the ease of transformation and regeneration, the annual yield of seed per hectare, the yield of recombinant protein per kilogram of seed, the producer price of the crop, the percentage of the seed that is made up of protein and, inevitably, intellectual property issues. Together, these determine the overall cost of producing the recombinant protein in the chosen seed crop.

Maize (Zea mays)

Maize was chosen as a platform expression host by ProdiGene Inc., College Station, TX, and has the honor of being the only crop thus far to be used commercially for the production of plant-derived recombinant proteins (avidin and β-glucuronidase were first produced in maize seeds on a commercial basis in 1997). Maize was chosen over the other cereals because it has the highest annual grain yield (8300 kg ha^{-1}), and the seeds have a relatively high seed protein content (10%) resulting in potentially the highest recombinant protein yields per hectare. Maize is also relatively easy to transform and manipulate in the laboratory, while in the field it has the greatest capacity of all the cereals for rapid scale up. Maize is the most widely cultivated crop in North America, so the complex infrastructure for growing, harvesting, processing, storing and transporting large volumes of corn is already in place. Prodigene is developing maize for the production of a range of pharmaceutical and technical proteins, including recombinant antibodies, vaccine candidates and enzymes.

Rice (Oryza sativa)

Rice is the most important staple food crop in the world and has also emerged as the model cereal species (it is the only terrestrial plant other than *Arabidopsis thaliana* to benefit from a completed genome sequence, and extensive EST resources are also available).

Rice has a lower annual grain yield than maize (6600 kg ha^{-1}) and the grain has a lower protein content (8%), but like maize it is easy to transform and manipulate in the laboratory, a range of useful expression cassettes have been developed, and field populations can be scaled up rapidly. The major disadvantage of rice compared to maize is that the producer price is significantly higher, so it may be excluded as a major expression host in the West simply on economic grounds. However, the story may be different in Asia and Africa, where rice is traditionally grown. Several pharmaceutical proteins have been expressed in rice grains, including α-interferon, recombinant antibodies and most recently a cedar pollen allergen. Rice suspension cell cultures have also been used for the production of pharmaceutical proteins.

Wheat (Triticum aestivium)

The main advantages of wheat for molecular farming are the low producer price and the high protein content of the grain (>12%). The adoption of this species as a production crop will depend, however, on the development of better transformation procedures and stronger expression cassettes, since at the current time only low expression levels have been achieved in transgenic wheat grains. The annual grain yield per hectare (2800 kg) is also much lower in wheat than in maize or rice, so larger areas of land would need to be cultivated to produce the same amounts of protein.

Barley (Hordeum vulgare)

Barley, like wheat, has a low producer price and a high protein content in the grain (about 13%). Also like wheat, gene transfer and regeneration techniques are not so well developed as they are in maize and rice. A significant advantage of barley, however, is that while few molecular farming experiments have been carried out using this species, the results have generally been very encouraging in terms of recombinant protein yields. In early reports, recombinant glucanase and xylanase were expressed at very low levels in barley. More recently, however, a recombinant diagnostic antibody was expressed in transgenic grains at levels exceeding 150 $\mu g\ g^{-1}$, and a recombinant cellulase enzyme was expressed at levels exceeding 1.5% total seed protein. Further recombinant proteins will need to be expressed in barley before its performance can be compared meaningfully with maize and rice.

Soybean (Glycine max)

The advantage of soybean as a production crop is that while it has a relatively low annual grain yield compared to maize and rice (about 2500 kg ha^{-1}), the protein content of the seed is very high (>40%) which means that it has the highest potential recombinant protein yields per hectare of any seed crop. This, combined with its relatively low producer price, makes it one of the least expensive crops for recombinant protein production. However, these advantages are balanced by the more difficult transformation and regeneration procedures. Therefore, only one report of molecular farming in soybean has been published, that of a humanized antibody against herpes simplex virus, and this was produced constitutively in the plant rather than in the seeds alone.

Although the use of soybean as a production host has been limited, the anti-HSV2 antibody produced in soybean is one of the few plant-derived antibodies in clinical development. Another potential advantage of soybeans is the ability to express proteins in the seed coat. This structure contains very few complex soluble proteins, which makes isolation of the target protein simple and inexpensive. Furthermore, the seed coats are easily removed and separated in the milling process, which makes them readily attainable and free of contamination from other components of the seed.

Pea (Pisum sativum)

Pea has a similar annual grain yield and seed protein content to soybean, and therefore has the same potential in terms of high recombinant protein yields per hectare. However, the producer price is about 50% higher than that of soybean, so proportionately higher yields would be necessary to make this a more competitive crop. To date, only a single pharmaceutical protein has been expressed in pea, a recombinant scFv antibody recognizing a cancer antigen. This was expressed under the control of the seed-specific legumin A promoter, and the maximum expression level achieved was 9 μg g^{-1} of seed. Trials have also been carried out with field pea varieties producing the enzyme ci-amylase from *Bacillus licheniformis* under the control of the bean USP (unknown seed protein) promoter.

Pigeon pea (Cajanus cajan)

Pigeon pea is often used in Africa, the Middle East and South Asia as a fodder crop, and is therefore suitable as an expression host for animal vaccines against diseases endemic in those areas. In one report thus far, pigeon pea plants have been generated expressing Rinderpest virus hemagglutinin at a level just below 0.5% TSP, with the aim of protecting ruminants against the disease caused by this virus. As with many crops envisaged as delivery vehicles for vaccines, this is a very specialized application and the crop is unlikely to be used as a general production system for recombinant proteins.

Peanut (Arachis hypogaea)

Peanuts are used as fodder in much of Africa and Asia, making this species also a suitable delivery vehicle for vaccines against animal diseases. As is the case for pigeon pea, transgenic peanut plants expressing Rinderpest virus hemagglutinin have been generated, but more general applications are unlikely.

Fruit and Vegetable Crops

Most fruit and vegetable crops that have been considered as hosts for molecular farming have been chosen not because they are particularly advantageous for bulk production, but because they might serve as vehicles for the delivery of edible vaccines. Therefore, the choice of a fruit or vegetable production crop usually reflects either geographical or cultural preferences that suit the production and distribution of a given vaccine, rather than any intrinsic advantages in terms of yields or stability that each species may exhibit. The exceptions to this general rule are potato and tomato, both of which have merits as general production platforms.

Potato (Solanum tuberosum)

The potato is the fourth most important food crop (after rice, wheat and maize) and is therefore widely grown throughout the world. It also has a very high tuber biomass yield (about 125 tonnes per hectare annually) which makes it eminently suitable for bulk protein production. Like cereals and grain legumes, potato plants have specialized storage organs (tubers) that are adapted for the accumulation of large amounts of protein. Targeting recombinant proteins to these organs therefore enhances protein stability in a manner similar to seed endosperm-specific expression in cereals. The first example of molecular farming in potato was the expression of human serum albumin in 1990, although in this case the protein was expressed in leaves. Potato leaves have also been used to express an industrial cellulase. Pharmaceutical proteins that have been expressed in potato tubers include human glutamic acid decarboxylase, human interferons, human interleukins as well as various diagnostic and therapeutic antibodies. Potato has also been used to express nutrition-enhancing proteins, such as human milk casein, and antibacterial lysozyme as potential additives to baby food. The most widespread use of this species, however, has been in the production of vaccine candidates. At least 10 different vaccine subunits have been expressed in potato, and in three cases thus far these have been used in clinical trials. Recently, potato was used simultaneously to express three vaccine antigens: cholera toxin B and A2 subunits, rotavirus enterotoxin and enterotoxigenic *Escherichia coli* fimbrial antigen for protection against several enteric diseases. The only disadvantage of using potatoes for the delivery of vaccines is that potatoes are cooked before eating in normal domestic settings, and heating may denature the recombinant proteins and render them inactive in terms of eliciting an appropriate immune response. This limitation does not apply to the other fruit and vegetable crops discussed below. A disadvantage of potatoes for the production of other pharmaceutical proteins, which need to be isolated and purified, is that the large tuber starch content may interfere with downstream processing.

Tomato (Lycopersi con esculentum)

Tomato plants have a high fruit biomass yield (about 60 tonnes per hectare per year) and offer other advantages in terms of containment, because they are grown in greenhouses. Tomato fruits have therefore been investigated as a general production system in molecular farming, and have been used to express one recombinant antibody (an scFv recognizing carcinoembryonic antigen) and several potentially pharmaceutical proteins (e.g. angiotensin-converting enzyme). As with potatoes, however, the most widespread use of tomato fruits thus far has been in the expression of vaccine candidates. The first such report involved the expression of rabies surface glycoprotein, which achieved the relatively high expression level of 1% TSP. Other vaccines that have been expressed in tomato include cholera toxin B subunit respiratory syncytial virus-F protein and, most recently, hepatitis E virus partial OFR2 and the B subunit of *E. coli* heat-labile enterotoxin. On the negative side, tomatoes must be chilled after harvest to keep the recombinant proteins stable, and the yields of recombinant proteins in tomatoes are generally relatively low at the present time. However, this could be addressed in some cases by transforming the fruit chromoplasts rather than the nuclear genome.

Carrot (Daucus carota)

The carrot taproot is a useful site for protein accumulation since this is both a natural storage organ and the edible portion of the plant. Unlike potato, carrots do not need to be cooked prior to consumption, so vaccines are more likely to remain in their native conformation.

Several antigens have already been expressed in carrot, including the *Mycobacterium tuberculosis* MPT64 protein, a diabetes-associated autoantigen (an isoform of glutamic acid decarboxylase) and various derivatives of measles hemagglutinin.

Banana (Musa spp.)

Vaccine production in transgenic plants began with the idea that transgenic bananas could be used to administer oral vaccines to adults (consuming whole fruits) and children (fed with banana paste or puree). Banana plants are cheap to grow and the fruits are a staple food source in many developing countries where vaccination campaigns are needed the most. One limitation of bananas, however, is that the transgenic plants take nearly two years to produce. Transgenic banana plants have been produced expressing vaccines against measles virus and hepatitis B virus. In order to prevent vaccine-containing bananas entering the food chain, attempts are in progress to introduce a marker gene that turns the banana flesh blue, enabling them to be distinguished from wild type fruits.

Oilcrops

Oilcrops are potentially advantageous for molecular farming because the oil bodies in developing seeds can trap the recombinant proteins and can be used to facilitate extraction and processing. Oil bodies are seed-specific organelles whose function is to accumulate triacylglycerides. Each oil body comprises a triacylglyceride core surrounded by a phospholipid membrane, which is peppered with oil-body-specific proteins termed oleosins. Recombinant protein can be trapped in the oil bodies by expressing them as oleosin-fusion proteins. A proprietary system developed by the biotechnology, can then be used to extract the oil bodies and isolate the recombinant protein by endoproteolytic cleavage. It may also be possible to exploit the natural oils normally extracted from such crops as byproducts, which can be used to offset production costs.

Rapeseed/Canola (Brassica napus)

Brassica napus is a widely grown crop used primarily for the production of oil, which is classed as either rapeseed oil or canola oil depending on its quality and content. The seeds of this plant also have a high protein content (25%), which makes them suitable for molecular farming. Transgenic rapeseed/canola plants can be produced relatively easily, and scale-up is rapid due to the large number of seeds produced by each plant. Although the overall biomass yields from this crop are comparatively low (about 1 tonne per hectare) significant cost savings can be achieved during down-stream processing steps by targeting recombinant proteins to the oil bodies and using these organelles to facilitate protein extraction and purification. This was first achieved in the case of leech hirudin by fusing the *Hirudo medicinalis* hirudin cDNA to the plant's oleosin gene.

The recombinant hirudin accumulated to 1% total seed protein, and following extraction of the oil bodies and purification of the fusion protein, the leech protein could be removed from its fusion partner by endoproteolytic cleavage in vitro. As well as providing a simple extraction and purification process, the expression of pharmaceutical proteins as fusions renders them inactive, and thus poses less risk to both the developing plant and any other organisms that come into adventitious contact with it. Enzymes have also been produced in rape-seeds, including a bacterial xylanase and *Aspergillus* phytase. Seeds expressing recombinant phytase are commercially available as a feed additive. One disadvantage of the oil body system is that proteins directed to oil bodies do not pass through the secretory pathway and therefore are not glycosylated. For this reason, the oil body system is not useful for the production of glycoproteins in cases where the N-glycans are necessary for function or activity.

Another useful feature of *B. napus* is the rapid sprouting of the seeds, which is the basis of a distinct platform technology being developed by UniCrop. In this method, proteins are expressed in the developing sprouts using cotyledon-specific Rubisco small subunit promoters, and proteins are extracted from the sprouts which are grown in an airlift tank. Although lower yields are produced in this method compared to agricultural scale production, the added advantage of containment may be useful for the production of pharmaceutical proteins under defined conditions.

Falseflax (Camelina sativa)

Falseflax is a self-pollinating oilseed crop, rarely used for food, which originates from the Fertile Crescent. This is also being developed as a sprout-based production platform by UniCrop.

Safflower (Carthamus tinctorius)

Safflower has many qualities in common with rapeseed including the suitability for oleosin fusion technology, and has been chosen as a platform crop by SemBioSys, for the following reasons: Safflower plants are readily transformed, they grow counter-seasonally, and they can be contained easily (there are no weedy relatives in the West, and seeds show minimal dormancy). The system is easily scalable and can produce clinical quantities of pure protein with an unprecedented manufacturing capacity. In addition, the seed-based system offers seasonally-independent availability of raw materials and improved inventory management since the recombinant proteins are stable in transgenic seeds for extended periods.

Unicellular Plants and Aquatic Plants Maintained in Bioreactors

Single-celled plants and aquatic plants can be maintained in bioreactors, which offers two critical advantages over molecular farming in terrestrial plants. First, the growth conditions can be controlled precisely, which means that optimal growth conditions can be maintained, batch-to-batch product consistency is improved and the growth cycle can conform to *good manufacturing practice* (GMP) procedures. Second, growth in bioreactors offers complete containment, thus sidestepping the environmental biosafety issues associated with all transgenic terrestrial plants, whether or not they are used for molecular farming. Although more expensive than agricultural molecular farming, the use of simple plants in bioreactors is not as expensive as cultured animal cells because the media requirements are generally very simple. Added to this, the proteins can be secreted into the medium, which reduces the downstream processing costs and allows the product to be collected in a non-destructive manner. A final, major advantage is the speed of production. The time from transformation to first product recovery is on the scale of days to weeks because no regeneration is required, and stable producer lines can be established in weeks rather than months to years because there is no need for crossing, seed-collection and the testing of several filial generations to check transgene stability. Three major bioreactor-based systems are currently under commercial development: algae, moss and duckweed.

Chlamydomonas reinhardtii

Although the alga *Chlamydomonas reinhardtii* is a model organism which has been instrumental in the study of photosynthesis and light-regulated gene expression, it has only recently been explored as a potential host for molecular farming. A single report discusses the production of monoclonal antibodies in algae, and shows that the production costs are similar to those of recombinant proteins produced in terrestrial plants, mainly due to the inexpensive media requirements (the medium does not cost very much to start with, and in any case can be recycled for algal cultures grown in continuous cycles). Aside from the economy of producing recombinant proteins in algae, there are further attributes that make alga ideal candidates for recombinant protein production. First, transgenic algae can be generated quickly, requiring only a few weeks between the generation of initial transformants and their scale up to production volumes. Second, both the chloroplast and nuclear genome of algae can be genetically transformed, providing scope for the production of several different proteins simultaneously. In addition, algae have the ability to be grown on various scales, ranging from a few milliliters to 500,000 liters in a cost-effective manner. These attributes, and the fact that green algae fall into the GRAS (generally regarded as safe) category, make *C. reinhardtii* a particularly attractive alternative to other plants for the expression of recombinant proteins.

Physcomitrella patens

The moss *Physcomitrella patens* is a haploid bryophyte which can be grown in bioreactors in the same way as algae, suspension cells and aquatic plants. Like these other systems, it has the advantages of controlled growth conditions, synthetic growth media, and the ability to secrete recombinant proteins into the medium. The unique feature of this organism, relative to all other plants, is that it is amenable to homologous recombination. This means that not only can it be transformed stably with new genetic information, but that endogenous genes can be disrupted by gene targeting. The major application of gene targeting in molecular farming is the modification of the glycosylation pathway (by knocking out enzymes that add non-human glycan chains to proteins) thus allowing the production of humanized glycoproteins.

The *P. patens* system is being developed by the German biotechnology company, which is based in Freiburg. The company has developed transient expression systems that allow feasibility studies, and stable production strains that can be scaled up to several thousand liters.

Duckweed (Lemna minor)

The *Lemna* System developed by the US biotechnology company Biolex Inc. has a number of significant advantages for the production of recombinant pharmaceutical proteins. Unlike transgenic terrestrial plants, this aquatic plant is cultured in sealed, aseptic vessels under constant growth conditions (temperature, pH and artificial light). Only very simple nutrients are required (water, air and completely synthetic inorganic salts) and under these conditions, the plant proliferates vegetatively and doubles its biomass every 36 hours. This provides the optimal production environment for batch-to-batch consistency. Duckweed constitutes about 30% dry weight of protein, and recombinant proteins can either be extracted from wet plant biomass or secreted into the growth medium. Biolex Inc. has reported the successful expression of 12 proteins in this system, including growth hormone, interferon-α, and several recombinant antibodies and enzymes.

Non-cultivated Model Plants

Rather than study every different species in the world, researchers focus on model organisms to learn general principles that can be applied to a wider range of life forms. Model organisms are often chosen for historical reasons, or because they are particularly easy to handle and manipulate in the laboratory. Model organisms may be chosen because they have short generation intervals or other features that make them amenable to genetic analysis. They may be chosen because they are particularly suitable for genetic manipulation or surgical procedures, or they may have a small, compact genome compared to related species. Often, a combination of the above features is present. Model organisms are rarely, if ever, chosen because of their commercial value. In terms of molecular farming, the advantage of model plants is that they are very well-characterized, which means that there is a disproportionately large amount of genetic, biochemical and physiological data available about them. This makes them suitable test systems for the design of novel expression systems. Three model plants, rice, tobacco and *C. reinhardtii*, have already been discussed in this section. The most important model plant of all, however, is *Arabidopsis thaliana*.

Arabidopsis thaliana

Arabidopsis thaliana is a small dicot plant of the mustard family. It has a number of features that make it ideal as a model organism: e.g. its size, short life cycle, the fact that it produces large numbers of seeds, and its relatively small genome of 125 Mb. The genome sequence of *A. thaliana* was completed in 2000, extensive EST resources are available and in terms of functional annotation, more is known about this higher plant than any other. The availability of such large amounts of information, together with the ease of transformation by methods such as floral dipping, make *A. thaliana* a convenient host species to use as a test system for molecular farming. Various pharmaceutical and industrial proteins have been expressed in the leaves, seeds and undifferentiated callus of this plant, including an *Acidothermus* endoglucanse which accumulated to 26% of *total soluble protein* (TSP).

Despite the encouraging results in terms of protein expression levels, however, *A. thaliana* is not suitable for cultivation on an agricultural scale. In this setting, its small size, weediness and low biomass yield are disadvantages and the plant has no commercial value. Therefore, while useful as a test system, it is unlikely this species will ever be used for commercial molecular farming.

Expression Systems for Molecular Farming

While the range of available expression hosts for molecular farming is impressive, the choice becomes even more varied when the different expression systems are considered. In this chapter, an expression host and an expression system are considered as separate entities: an *expression host* is defined as a particular species while an *expression system* is defined as a transformation, propagation and expression strategy. The combination of host and system results in an *expression platform*, e.g. stably transformed rice seeds, transiently transformed alfalfa leaves, stably transformed tobacco suspension cells, virus-infected spinach leaves etc. Note that this is not a universal nomenclature, and in the literature the terms host, system, platform, strategy etc. appear to be used interchangeably. By far the predominant system used in molecular farming is the nuclear transgenic plant with the recombinant protein accumulating within the plant tissues. However, biosafety concerns and the disadvantage of long development phases have driven researchers to look at alternative systems based either on transient expression, extra-nuclear transgenesis, or the use of cultured plants, organs or cells.

Transgenic Plants

Transgenic plants usually contain foreign DNA incorporated into the nuclear genome. Such plants have a combination of valuable attributes for molecular farming, but in terms of commercialization the most important of these are the low overall cost of production, the inherent scalability of agricultural systems, the fact that stable transgenic lines are a permanent resource, and the variety of production hosts suitable for different applications. Scalability is probably the most important advantage because the cost of recombinant proteins produced in field plants is inversely proportional to the production scale. In a market which can see demand rapidly increase and decrease, fermenter-based production systems are often unable to cope or left with surplus capacity. In contrast, the scale of plant-based production can be modulated rapidly simply by using more or less land as required. It can take several years to achieve e.g. a tenfold scale-up or scale-down in fermenter systems or in transgenic animals, but a field of transgenic plants can be scaled up or down more than 1000-fold in a single generation by planting greater or smaller numbers of seeds.

The two major disadvantages of transgenic plants are the development timescales and the increasing importance of biosafety and regulatory compliance. The '*gene-to-protein*' time for transgenic plants encompasses the preparation of expression constructs, transformation, regeneration and the production and testing of several generations of plants. The testing phase is necessary to ensure transgene and expression stability and the biochemical activity of the product, as well as the absence of adverse phenotypic changes in the host plant. These processes take up to two years depending on the plant species although milligram amounts of protein might be available after several months for initial testing. Biosafety concerns include the risk of transgenic plants becoming naturalized outside their intended production sites due to human or animal activities, the risk oftransgene spread by outcrossing or horizontal gene transfer, and the risk to human health and the environment caused by the presence of potentially toxic recombinant proteins. One further limitation is the low yields that are obtained in many transgenic plants. In some cases, this reflects the poor performance of the expression construct and can be addressed by improved construct design. However, in other cases the problem is intrinsic to the transgenic plant, and reflects rate limiting processes oftransgene expression, protein synthesis or protein turnover.

Transplastomic Plants

Transplastomic plants are transgenic plants generated by introducing DNA into the chloroplast genome, usually by particle bombardment. The plants are grown in the same way as nuclear transgenic plants and therefore suffer the same disadvantages in terms of production timelines. However, the advantages of chloroplast transformation are many: the transgene copy number is high because of the many chloroplasts in a typical photosynthetic cell, there is no gene silencing, multiple genes can be expressed in operons, the recombinant proteins accumulate within the chloroplast thus limiting toxicity to the host plant, and the absence of functional chloroplast DNA in the pollen of most crops provides natural transgene containment. The high transgene copy numbers and the absence of silencing have resulted in extraordinary expression levels, e.g. 25% TSP for a tetanus toxin fragment, 11% TSP for human serum albumin and 6% TSP for a thermostable xylanase. The ability to express multiple genes in operons means that chloroplast expression will be particularly amenable to the production of multi-subunit proteins such as antibodies. However, while proteins expressed in the chloroplast have been shown to fold properly and form appropriate disulfide bonds, glycosylation does not occur, so this system has limited use for the production of therapeutic glycoproteins. The biosafety advantages of the chloroplast system are particularly noteworthy since the inability of functional chloroplast DNA to reach the egg in most commercial crop species means that transgene spread by outcrossing is strongly inhibited. However, other biosafety concerns such as seed spread by human and animal activities are just as prevalent in transplastomic crops as they are for standard transgenic crops. At the current time, chloroplast transformation is a routine procedure only in tobacco and *C. reinhardtii*. However, plastid transformation has been achieved in a growing number of other species, including carrot and tomato. The ability to transform the chromoplasts of fruit and vegetable crops has obvious advantages for the expression of subunit vaccines.

Virus-infected Plants

Recombinant plant viruses have been used as expression vectors because the infections are rapid and systemic, leading to high levels of recombinant protein production soon after inoculation. Compared to transgenic and transplastomic plants, the development phase is significantly reduced, but the scalability of virus-mediated expression is equivalent to that of other field plant expression systems. No known plant viruses integrate into the genome, so the genetic modification of plants is entirely avoided. Two major strategies have been developed with viral vectors: the expression of full-length recombinant proteins and the presentation of foreign epitopes on the surface of viral particles. Both TMV and PVX have been used in the context of the first strategy to produce antibodies, vaccine candidates and some other pharmaceutical proteins. *Cowpea mosaic virus* (CPMV) and AlMV are the most popular epitope presentation systems, but TMV and PVX have also been used for this purpose.

Perhaps the most significant use of viral expression systems for molecular farming was described by McCormick and colleagues. These investigators used TMV vectors in *N. benthamiana* to produce a scFv antibody based on the idiotype of malignant B-cells from the murine 38C13 B-lymphoma cell line. When administered to mice, the recombinant protein stimulated the production of anti-idiotype antibodies capable of recognizing 38C13 cells, providing immunity against lethal challenge with the lymphoma. This strategy could be used to develop personalized therapies for diseases such as non-Hodgkin's lymphoma. Antibodies capable of recognizing unique markers on the surface of any malignant B-cells could be produced for each patient. The necessity for speed in the derivation of such prophylactic antibodies is recognized by the use of viral vectors rather than transgenic plants. These antibodies are now undergoing Phase I clinical trials. Another important development in the use of viral vectors was the production of full-size immunoglobulins in *N. benthamiana* plants using two TMV vectors. One of the vectors expressed the heavy chain and one the light chain. This study showed that viral coexpression was compatible with the correct assembly and processing of multimeric recombinant proteins. Many epitopes from human and animal pathogenic viruses and bacteria have been expressed as coat protein fusions in CPMV and AlMV, and in most cases mice either injected with plant extracts or administered the extracts intranasally developed suitable immune responses.

Transiently Transformed Leaves

Transient expression assays are often used to evaluate expression constructs or test the functionality of a recombinant protein before committing to the long term goal of transgenic plants. However, transient expression can also be used as a routine molecular farming method if enough protein can be produced to make the system economically viable. An example of a transient expression system is the agroinfiltration method, where recombinant *Agrobacterium tumefaci ens* are infiltrated into leaf tissue and genes carried on the T-DNA are expressed for 2–5 days without integration. Agroinfiltration was developed in tobacco, but there appears to be no intrinsic limitation to the range of species that can be used, since preliminary data obtained at the RWTH Aachen demonstrates the successful infiltration of over 20 different plant species. Although originally considered difficult to scale up, agroinfiltration is now known to be suitable for the routine production of milligram amounts of protein in a timescale of weeks. Scientists at Medicago Inc., regularly process up to 7500 infiltrated alfalfa leaves per week and similarly we have shown that up to 100 kg of wild type tobacco leaves can be processed by agroinfiltration, resulting in the production of 50–150 mg of protein per kg.

A number of different antibodies and their derivatives have been produced by agroinfiltration, including the full-size IgG T84.66 (along with its scFv and diabody derivatives), and a chimeric full-size IgG known as PIPP which recognizes human chorionic gonadotropin. Recently, several reports have described how agroinfiltration can be scaled-up more efficiently. Baulcombe and colleagues have shown that the loss of protein expression seen a few days after agroinfiltration is predominantly caused by gene silencing. They managed to increase the expression levels of several proteins at least 50-fold by co-expressing the p19 protein from tomato bushy stunt virus, a known inhibitor of gene silencing.

Hydroponic Cultures

In most cases where nuclear transgenic plants have been used for the production of recombinant protein, the proteins have been extracted from plant tissues. An alternative is to attach a signal peptide to the recombinant protein thus directing it to the secretory pathway. In this way, the protein can be recovered from the root exudates or leaf guttation fluid, processes known respectively as rhizosecretion and phyllosecretion. Although not widely used, the secretion of recombinant proteins into hydroponic culture medium is advantageous because no cropping or harvesting is necessary. In an exciting recent development, a monoclonal antibody was shown to be secreted into hydroponic culture medium resulting in a yield of 11.7 μg antibody per gram of dry root mass per day.

Hairy Roots

Hairy roots are neoplastic structures that arise following transformation of a suitable plant host with *Agrobacterium rhizogenes*. If the plant is already transgenic, or if the transforming *A. rhizogenes* strain is transgenic and transfers the foreign gene to the host plant during the process of transformation, then hairy root cultures can be initiated which will produce recombinant proteins and secrete them into the growth medium. Hairy roots grow rapidly and can be propagated indefinitely in liquid medium. Thus far, hairy root cultures have been used to produce a relatively small number of antibodies mainly because of the relative ease with which multi-subunit proteins can be produced. The cultures can be initiated from transgenic plants already carrying multiple transgenes, wild type plants can be infected with multiple *A. rhizogenes* strains, or established hairy root cultures can be super-transformed with *A. tumefaciens*.

Shooty Teratomas

Shooty teratomas are differentiated cell cultures produced by transformation with certain strains of *A. tumefaciens*. Thus far, there has been only one report of pharmaceutical protein production in teratoma cultures, and the levels of antibody were very low.

Suspension Cell Cultures

Suspension cell cultures are usually derived from callus tissue by the disaggregation of friable callus pieces in shake bottles or fermenters of liquid medium. Recombinant protein production is achieved by using transgenic explants to derive the cultures, or transforming the cells after disaggregation, usually by co-cultivation with *A. tumefaciens*. Suspension cultures have the same advantages as the simple plants, i.e. controlled growth conditions, batch-to-batch reproducibility, containment and production under GMP procedures. The main disadvantage is the scale of production, although tobacco suspension cells have been cultivated at volumes of up to 100,000 liters. Many foreign proteins have been expressed successfully in suspension cells, including antibodies, enzymes, cytokines and hormones. Tobacco cultivar Bright Yellow 2 (BY-2) is the most popular source of suspension cells for molecular farming, since these proliferate rapidly and are easy to transform. However, rice suspension cells have been used to produce several biopharmaceutical proteins and soybean suspension cells have been used to produce a hepatitis B vaccine candidate. Proteins expressed in suspension cells can either be extracted from the wet biomass or secreted into the culture medium for continuous, non-destructive recovery.

Expression Strategies and Protein Yields

Finally, we consider some general strategies for the control of gene expression and protein accumulation in plants. These strategies play an important role in defining the overall yields of recombinant proteins, but they also have wider impact, e.g. on biosafety and product authenticity. To achieve high yields, all stages of gene expression must be optimized, including transcription, mRNA stability, mRNA processing, protein synthesis, protein modification, protein accumulation and protein stability. Since expression constructs are chimeric structures in which the transgene is enclosed by various regulatory elements, the considered choice of these regulatory elements is an essential component of the development phase in molecular farming.

For high-level transcription, the two most important elements are the promoter and the polyadenylation site. In dicot plants, the *cauliflower mosaic virus* (CaMV) 35S promoter is the most popular choice because it is strong and constitutive, and therefore drives high-level transgene expression in leaves, fruits, tubers, roots and any other relevant organs. The promoter can be made even more active by various modifications, such as duplicating the enhancer region. In monocots, where seed expression is the normal strategy, the CaMV 35S promoter has lower activity and is generally replaced either with the maize ubiquitin-1 promoter (which is constitutive, but drives high level transgene expression in seeds), or with a seed-specific promoter from a seed storage protein gene (e.g. maize zein, rice glutelin, bean *unknown seed protein* (USP), pea legumin). The seed-specific *arc5-I* promoter from the common bean (*Phaseolus vulgaris*) has been used to express a single chain antibody in *Arabidopsis thaliana*, and this resulted in the accumulation of 36 times more recombinant protein than when the transgene was driven by the CaMV 35S promoter. Other tissue-specific promoters that are useful in molecular farming include fruit specific promoters for tomato and tuber-specific promoters for potato. In each case, restriction of transgene expression to the target tissue prevents expression in vegetative organs, therefore reducing any negative impact of the recombinant protein on normal plant growth and development, and limiting the exposure of non-target organisms such as pollinating insects or microbes in the rhizosphere.

Inducible promoter systems are also valuable assets in molecular farming because transgene expression can be controlled externally. One example is the *mechanical gene activation* (MeGA) system. This utilizes a tomato hydroxy-3-methylglutaryl CoA reductase 2 (HMGR2) promoter, which is inducible by mechanical stress. Transgene expression is activated when harvested tobacco leaves are sheared during processing, which leads to the rapid induction of protein expression, usually within 24 hours. Another potentially useful inducible promoter that has been described recently is the peroxidase gene promoter from sweet potato (*Ipomoea batatas*). When linked to the *gus*A reporter gene, this promoter produced 30 times more GUS activity than the CaMV 35S promoter following exposure to hydrogen peroxide, wounding or ultraviolet light.

Other parts of the expression construct are also important. A strong polyadenylation signal is required for transcript stability and those from the CaMV 35S transcript, the *Agrobacterium tumefaciens nos* gene and the pea *ssu* gene are popular choices. In monocots, the presence of an intron in the 5' untranslated region of the expression construct has been shown to improve transgene expression. The structure of the 5' and 3' untranslated regions should also be inspected for AU-rich sequences, which can act as cryptic splice sites and instability elements. Some sequences, such as the 5' leader of the petunia chalcone synthase gene, have been identified as translational enhancers and these can be incorporated into the expression construct to boost protein synthesis.

Other important factors that influence translation include the presence of a consensus Kozak sequence, the absence of multiple AUG codons, and the disparity of codon bias between the transgene donor and the host species.

One of the most important considerations for the improvement of protein yields is subcellular protein targeting, because the compartment in which a recombinant protein accumulates strongly influences the interrelated processes of folding, assembly and post-translational modification. All of these contribute to protein stability and hence help to determine the final yield.

Comparative targeting experiments with full size immunoglobulins and single chain Fv fragments have shown that the secretory pathway is often a more suitable compartment for folding and assembly than the cytosol, and is therefore advantageous for high-level protein accumulation. Because many plant-derived recombinant proteins under development are human proteins that normally pass through the endomembrane system, this principle can be applied not only to antibodies but also more generally. However, it is not a universal rule, since there are examples of proteins that are more abundant when directed to the cytosol than the secretory pathway, e.g. α-galactosidase A. Antibodies targeted to the secretory pathway using either plant or animal N-terminal signal peptides usually accumulate to levels that are several orders of magnitude greater than those of antibodies expressed in the cytosol. Even with antibodies, however, there are occasional exceptions, and this suggests that intrinsic features of each antibody might also contribute to overall stability. The endoplasmic reticulum provides an oxidizing environment and an abundance of molecular chaperones, while there are few proteases. These are likely to be the most important factors affecting protein folding and assembly. It has been shown recently that antibodies targeted to the secretory pathway in transgenic plants interact specifically with the molecular chaperone BiP.

In the absence of further targeting information, the expressed protein is secreted to the apoplast. The stability of antibodies in the apoplast is lower than in the lumen of the ER. Therefore, antibody expression levels can be increased up to ten times higher if the protein is retrieved to the ER lumen using an H/KDEL C-terminal tetrapeptide tag. Again, although the principles of ER-retention in molecular farming have been established using antibodies, it is likely that they will also apply to many other proteins.

In this chapter, we have looked at the properties of different expression hosts and expression systems, and considered some of the available strategies to control transgene expression and protein accumulation. When all these variations are combined, there exists a very diverse range of potential expression platforms that can be used to produce recombinant proteins. The choice depends on many factors, some intrinsic to the plant species or expression system, some dependent on the recombinant protein and its intended use, and some determined by external factors such as regional, economic and regulatory constraints. There is no ideal production platform for molecular farming, and each of the host plants and systems described in this chapter has its merits and drawbacks. As ever, the choice of production platform therefore should be determined empirically, and on a case-by-case basis.

10

Energetic System

Electrical power systems that serve pharmaceutical equipment must be safe, reliable, functional, predictable, flexible, clean, and sometimes validated. The electrical power systems have voltages ranging from 120 to 69,000 V. Pharmaceutical plants in the United States and Canada can purchase 3 phase, 60 Hz electric power from utility companies that is more reliable than the power they could generate in house. Whereas in other countries, electric power is purchased at 3 phase, 50 Hz, and may not be as reliable as power that is generated in the plant.

This article is meant to convey a basic understanding of various power distribution system configurations and familiarize the reader with the electrical distribution equipment that are commonly installed within pharmaceutical plants in the United States. It will help in the evaluation of both new and existing electrical power distribution systems that serve pharmaceutical equipment.

Electrical Power Sources

Electricity can be purchased from an energy provider or can be generated in the pharmaceutical plant. Usually, purchased electricity is cheaper than on-site generated electricity. Large plants have many options when purchasing electricity. A thorough understanding of these options can significantly reduce the electrical operating cost and will increase the reliability of the electrical power system. All plants and especially small- and medium-size plants must accurately determine their current and future normal electrical power loads and normal/ emergency power loads when deciding how to purchase electricity and how to install a reliable electrical power system.

The pharmaceutical plant's electrical loads can be divided into four categories:

1. *Normal loads*. Loads that can be turned off for a period of time without creating a hazardous condition or causing a substantial loss of product or research.
2. *Standby loads*. Loads that if they lose electricity will cause a hazardous condition or cause a substantial loss of product or research but which can sustain a power interruption of 60 sec.
3. Standby non-interruptible load—Loads that cannot sustain any interruption in power.
4. Emergency loads—Legally required emergency and egress lighting and other loads classified as such by governmental agencies and locally adopted building codes.

Loads can be further divided into groups based on their utilization voltage, category, type, and location within the plant. Typical utilization voltages within pharmaceutical plants include 480/277 and 208/120 V. Categories include linear and non-linear loads. Linear loads typically include incandescent lamps and induction motors that are not controlled by a *variable frequency drive* (VFD). Non-linear loads include all loads that have switching power supplies or silicon control rectifiers such as VFDs, uninterruptible power supplies, and electronic ballasts on 480/277 V systems and personal computers, copy machines, faxes, electronic devices, and instruments on 208/120 V systems. Types include continuous loads such as lighting that stay on all day and non-continuous loads that will not operate for more than 3 hr at a time. The areas in the plant should be divided into locations based on the function or process performed in the area and based on whether or not the area being classified is a current good manufacturing practices (cGMP) area.

The load projections should be based on a minimum of 7 years. To obtain the group demand for each group of loads, multiply the connected load for the group by a diversity factor that reflects the proportion of load that will operate at any one time to the total connected load. The sum of the group demands will equal the plant's maximum demand. After determining the maximum demand, consult with the local public utility company for assistance in determining the optimum voltage and configuration for the electrical service entrance for the plant.

The ampacity or rated current carrying capacity of the electrical service entrance conductors that connect the utility company's lines to the plant's service entrance equipment must be a minimum of 125% of the calculated maximum demand for continuous loads plus 100% of the maximum calculated demand for non-continuous loads. Service entrance conductors and equipment with higher ratings or provisions to increase the rating of the service entrance conductors are recommended.

Large plants, which have a *central utility plant* (CUP) with chiller motors rated above 200 hp, should consider purchasing and in some instances distributing electricity at primary voltage levels such as 4160 V and 13,800 V.

An emergency generator should have a rating that is 25% and possibly 100% higher than the sum of the total calculated standby and emergency demand for the plant. Alternately, provisions for a future emergency generator should be provided as part of the initial installation. Discuss the type of fuel required for the generator's engine with the local authority having jurisdiction. The National Electrical Code, NFPA 70 (NEC), requires that the fuel be on site if the generator provides power for emergency loads. However, exceptions to this requirement have been permitted where reliable natural gas service is available.

Electrical Power System Utility Services

Primary services are typically installed for medium and large pharmaceutical plants and can be purchased as a single service, a dual service, or a regular-reserve service. The utility company will install two separate power lines to the plant for a dual service and for a regular-reserve service. Each dual service power line will normally serve half of the loads in the plant but each will be capable of serving all of the loads in the plant. Regular-reserve service power lines are connected to a common bus with only the regular power line normally serving all of the plant's loads.

Secondary services are typically installed for small- to medium-size pharmaceutical plants. They are available at the utilization voltages of 277/480 or 120/208 V. Secondary services are less expensive to install and maintain than primary services. However, the kilowatt-hour cost for electrical power supplied by a secondary service is higher.

Secondary Configurations

The following items should be considered when planning an electrical power system's secondary, less than 1000 V, configuration.

1. The utility company will provide and own the transformer that will supply secondary voltage for the service to the plant. This transformer should be dedicated to the pharmaceutical plant and

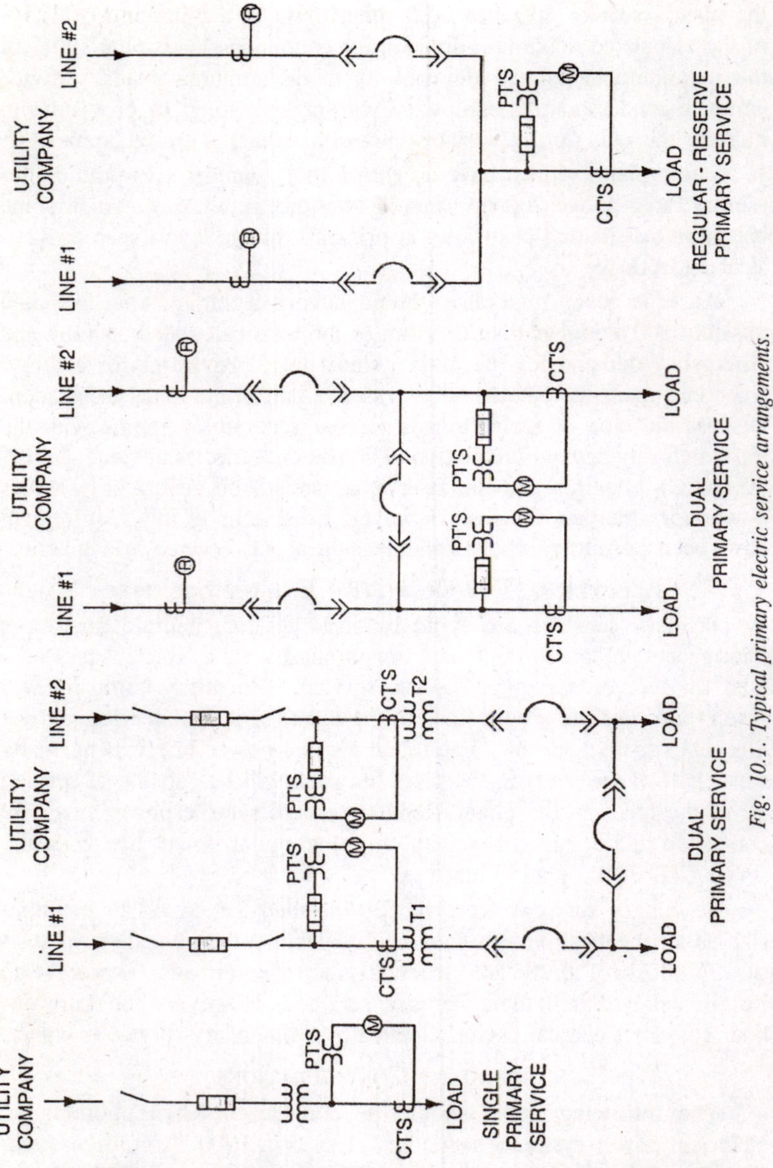

Fig. 10.1. Typical primary electric service arrangements.

should have a grounded wye secondary. Sharing a utility-owned transformer with other utility customers will reduce the quality and the reliability of the electric service to the plant.

2. The size of the utility company's transformer will affect the rating of the service entrance equipment and the starting of large motors. If the transformer rating is low, reduced voltage starting will be required for starting large motors. If the transformer rating is high, service entrance equipment will require a high symmetrical short-circuit (IAC) rating. The utility company will provide the maximum available symmetrical short-circuit current available at the secondary of their transformer.
3. Service entrance equipment must have a current rating equal to or greater than the present and future service entrance conductors' ampacity. Moreover, this equipment must have an IAC current rating that is higher than the available short circuit current at the utility company's transformer. Spare service entrance conduits should be installed for future service entrance conductors. The service entrance equipment should include spaces for future feeder circuit breakers to supply future loads. All incoming service entrance conductors must be 3 phase, 4 wire and should be installed in underground metal conduit.
4. Service entrance equipment should be installed in an indoor electrical room whenever possible. The equipment should include one to six main service entrance power circuit breakers or load break disconnect switches and feeder circuit breakers for loads within the plant. Circuit breakers and disconnect switches for 480/277 V systems that are rated at 1000 A or more must include ground fault protection. Feeder circuit breakers will typically be molded case. Integrally fused circuit breakers are available with very high short circuit IAC ratings for both power and molded case circuit breakers. Fuses and current limiters for integrally fused circuit breakers must be stored within the electrical room.
5. Fuses should not be used to protect secondary voltage feeders. The time current characteristics of fuses above 100 A will not coordinate with the ground fault pickup currents and time delays of the main overcurrent protection (circuit breaker or fused disconnect switch) ground fault protection. A main load break disconnect switch can be equipped with current-limiting fuses to reduce the available short-circuit current from the utility and should have a three-phase voltage relay for single-phase protection.
6. Ground fault protection, including circuit breakers or relays, should be set as low as possible without causing nuisance tripping. The ground fault pickup and time delay for the main overcurrent

protection should coordinate with the trip characteristics of the largest feeder circuit breaker that does not have ground fault protection and must be less than the maximum current permitted by the NEC.

7. If fuses are used in the power distribution system, three-phase motor starters should have solid-state overload devices that include single-phase protection.
8. Two (2) to six (6) separate service entrance cables connected to the secondary side of one utility owned transformer can connect to separate main circuit breakers or main load brake fused disconnect switches that are mounted next to each other. The sum of the ampacity of all of the service entrance conductors must have a current rating equal to or greater than the present and future maximum demand for the plant. This approach can reduce the cost of the service entrance installation but will not provide the additional level of protection provided by having one main circuit breaker or one main fused disconnect switch. Therefore it is not as reliable.
9. A maintenance program should document all tests, inspections, and faults cleared by a main and feeder circuit breaker. Lack of such a program will reduce the reliability of the electrical equipment. Molded case circuit breakers require no internal maintenance. They should be inspected for broken casing and loose connections after they operate to clear a fault and should be replaced after they have operated to clear two faults. Feeder circuit breakers, especially older circuit breakers, should be exercised (repeatedly opened and closed) whenever possible. Part of the initial commissioning of an electrical power system must include testing of the ground fault protection.
10. The secondary selective circuit arrangement is recommended for all critical loads. It requires the installation of double-ended switchgear or a variation thereof. Double-ended switchgear includes two main circuit breakers that serve separate feeder buses with a tie circuit breaker in the middle. Each feeder bus has its own set of feeder or branch circuit breakers. The load should be evenly split between each of the feeder buses.
11. Important loads can be connected to redundant feeder circuit breakers, one on each side of the double-ended switchgear, by connecting the load sides of the circuit breakers together with a feeder cable. This approach should be used with caution. Another

approach is to connect the redundant feeder circuit breakers to a time-delayed automatic transfer switch.

Primary Configurations

The following items should be considered when planning an electrical power distribution system's primary configuration.

1. A single primary service can have one to six primary service entrance circuit breakers or fused disconnect switches. These overcurrent devices will protect primary feeders that distribute power to distribution unit substations located in electrical rooms throughout the plant, to primary voltage motor starters in the CUP, and sometimes to generator switchgear when the generator system is designed for momentary or continuous operation in parallel with the utility line.
2. A dual service will have two main service entrance circuit breakers and a service entrance tie circuit breaker and may have additional primary feeder circuit breakers. A regular-reserve service will have two main service entrance circuit breakers and may have additional primary feeder circuit breakers. In very large pharmaceutical plants, the primary feeder circuit breakers will connect to outdoor substation transformers that will provide 4160 V for the CUP and 13,800 V for distribution unit substations.
3. Unit substations, each consisting of a primary fused disconnect switch or circuit breaker, a dry transformer, and secondary switchgear should be in a central location near their loads. Unit substation transformers should be sized to carry at least 125% the maximum present and future connected load.
4. Double-ended unit substations provide economical, reliable, preengineered, and factory-tested configuration for obtaining secondary selective circuit arrangements at utilization voltage of 480/277 or 120/208 V and for installing a primary selective circuit arrangements. Double-ended unit substations include one or two main primary circuit breakers or fused disconnect switches on both sides that are connected to transformers that are connected to the feeder circuit breaker sections. Each feeder circuit breaker section has a main secondary circuit breaker that protects the internal bus, which connects to the feeder circuit breaker, and is connected to the other feeder circuit breaker section through a secondary tie circuit breaker. Each transformer should be rated to carry at least 125% of the maximum present and future connected load for the entire double-ended substation.

5. Primary selective circuit arrangements utilize two primary selectable feeders for each transformer. These feeders are either connected to the line side of two interlocked primary non-fused disconnect switches that have a common load side fuse or to the line side of two interlocked primary fused disconnect switches. Only one primary disconnect switch can be closed at any time.
6. Dry-type transformers are typically installed indoors and include ventilated, sealed, or gas filled, totally enclosed non-ventilated, and cast coil types. Cast coil transformers are the most expensive and they have the highest overload capability. Transformers should be sized based on their air to air (AA) rating. Provisions for future automatic fans that will provide a higher force air (FA) rating and an alarm that indicates when the fans are operating due to the temperature within the transformer should be provided.
7. If installed outdoors, pad mounted liquid filled transformers up to 2500 KVA are available with mineral oil or less flammable silicon liquids. Outdoor liquid-filled transformers sometimes have an advantage over outdoor dry-type transformers, which present a relatively difficult lightning protection problem because of their lower basic impulse levels (insulation level). Substation transformers with ratings above 2500 KVA are utilized for higher secondary voltages, including 4160V for large motors and 13,800 V for medium voltage feeders.
8. All configurations described herein are based on radial feeders. Radial feeders provide power to feeders for panelboards and other electrical distribution equipment within the pharmaceutical plant. Branch circuits connect to feeders and provide power to the electrical loads.

Feeders and Branch Circuit Arrangements

Feeders originating in switchgear, panelboards, and motor control centers provide power to electrical distribution equipment within the pharmaceutical plant. Branch circuits originating in panelboards and motor control centers provide power to pharmaceutical equipment, lighting, motors, and other equipment.

1. Three phase, 3 wire or 3 phase, 4 wire power and lighting panelboards having current ratings from 100–1600 A may have a main circuit breaker or may have main lugs only. They will have feeder circuit breakers for other electrical equipment and/or branch circuit breakers for pharmaceutical equipment, lighting fixtures, etc. The panelboards should have an IAC rating greater than the

available short-circuit current calculated at the transformer serving the panel- board. Four-wire switchgear, panelboards, and motor control centers should have a separate neutral bus and a separate ground bus. The secondary neutral of a separately derived 4 wire system should be bonded to ground at one point ahead of the main secondary circuit breaker, ideally in the transformer and not in the panel-board. Panelboard schedules and nameplates must be installed and maintained for reliability and to comply with the NEC.

2. Separately derived systems that originate at the secondary of two winding transformers including single-phase and three-phase delta-wye transformers and at the output of a wye connected generators should be connected to the power distribution system with a 4 pole *automatic transfer switch* (ATS) when the neutral conductor is installed in the feeder to the ATS. This will avoid circulating ground currents that can trip ground fault relays and effect sensitive electronic equipment.
3. Secondary unit substations (120/208 V) and transformers need to be distributed throughout the pharmaceutical plant to provide 120 and 208 V branch circuits to pharmaceutical equipment. Because of their lower voltage, the length of these branch circuits should be short.
4. AC motors 1 hp and larger should be 3 phase, 460 V. Smaller motors can be 120V, single phase. If the plant purchases power at 208/ 120V, 3 phase or 240V, 3 phase, all motors 1 hp and larger should be 200 or 230 V, 3 phase motors.
5. Motor control centers (MCCs) are typically 480 V, 3 phase, 3 wire or 277/480 V, 3 phase, 4 wire with current ratings from 600–2500 A. Motor control centers should be a type that have been designed for industrial applications rather than commercial applications and must have IAC ratings greater than the available short-circuit current calculated at the transformers serving the MCCs. New MCCs that are designed to incorporate intelligent starters, protective equipment, network communication, and 24 V control circuits should be considered for future pharmaceutical process applications.
6. Individual motor controllers are sometimes connected to the main switchgear when they serve large motors but are more often connected to motor branch circuits originating in the distribution panelboards or MCCs. Individual motor controllers can be full

voltage non-reversing or full voltage reversing, reduced voltage, and solid state. Reduced voltage and solid state motor controllers are used when the load requires a soft start or when there is insufficient let through (short circuit) current to start the motor without causing a severe voltage drop. When controlling a piece of critical pharmaceutical equipment within a cGMP area or process, the operation of the motor controller and all of its control circuits should be part of the qualification and validation process.

7. Variable frequency drives are special types of motor controllers. They are almost always used with 480 V motors and sometimes for motors greater than 200hp because they require less line current to deliver the same amount of torque as compared to reduced voltage starters. They are complex and require a clean, cool, ventilated environment to properly operate. Variable frequency drives vary the speed of motors, which if continuously run at a slow speed may need supplemental cooling. Motors, designed to be controlled by VFDs, should be specified. Variable frequency drives generate harmonics on their input (line side) and output (load side) motor branch circuits. Capacitors cannot be installed on the load side of VFDs. Capacitors installed on the line side of VFDs at the switchgear or MCC must have a KVAR rating that will not cause a resonant circuit at any of the harmonic current frequencies generated by the VFDs. Variable frequency drives require many parameter settings, most of which can remain at the factory default settings and all of which should be recorded. When controlling a piece of critical pharmaceutical equipment within a cGMP area or process, the operation of the VFD, its parameter settings, and all of its control circuits should be part of the qualification and validation process.
8. Custom-built control cabinets are provided for various types of pharmaceutical equipment, usually by the equipment manufacturer. These control cabinets should be UL listed or labeled, otherwise all of the wire and electrical equipment in them are subject to the requirements of the NEC. The functions controlled by these control cabinets and sometimes the equipment within the control cabinets and all of its control circuits should be part of the qualification and validation process.
9. Uninterruptible power supplies (UPSs) must be installed for all standby non-interruptible loads. Smaller UPSs that are dedicated to individual pieces of pharmaceutical equipment are usually better

than a single large UPS. Because of the amount of non-linear loads within pharmaceutical distribution systems, UPSs are the only source of clean electricity that does not contain harmonics generated by other non-linear loads. Uninterruptible power supplies usually have parameter settings and alarms. When providing power to a piece of critical pharmaceutical equipment within a cGMP area or process, the operation of the UPS, its parameter settings, and alarms should be part of the qualification and validation process for the piece of equipment.

10. Automatic transfer switches (ATSs) are commonly used to transfer emergency and standby loads from a normal source of power to an emergency source of power. Automatic transfer switches are available at different voltages, with three or four poles, with contacts to start an emergency generator, with time delays when part of a priority load shedding circuit, with controls that will permit bumpless transfer between the normal and emergency power system, and with integral battery chargers for the emergency generator starting batteries. They can be simple or complex and are critical parts of a reliable power distribution system. Automatic transfer switches are a latent single point of failure within normal/ emergency circuits and therefore must be inspected and tested as often as possible.

Generators and Emergency Power Sources

A pharmaceutical plant can obtain emergency or standby electrical power from generators, rechargeable batteries, and in rare cases from a separate utility service. Uninterruptible power supplies, central storage battery system, and unit equipment all use rechargeable batteries for their emergency source of power.

1. Unit equipment (self-contained battery packs with integral or remote lamps) provide the lowest initial cost for emergency egress lighting and exit signs but require considerable maintenance.
2. Emergency generators can provide emergency power for egress lighting and exit signs and standby loads. The 1996 NEC introduced a requirement that all legally required emergency loads must be supplied by a dedicated ATS with dedicated feeders and branch circuits. Therefore installations after 1996 must have two or more ATSs when serving both emergency and standby loads. Emergency generators typically have a standby rating and sometimes have a lower prime rating. The standby rating should be used unless the generator is intended to run for weeks at a time. Standby generators

can deliver 100% of their rated KW output for short periods of time to linear loads. Non-linear loads will reduce the rating of the emergency generator. Outdoor emergency generators can be purchased with subbase double-wall fuel tanks and can be installed in walk-in enclosures in sizes up to 1600 KW.

When the entire pharmaceutical plant or a large portion thereof requires emergency power, two or more generators are recommended. These generators should have a priority load shedding circuit that turns off the non-critical loads before critical and legally required loads. Emergency generators that can be synchronized with each other can also synchronize with one of the utility lines if approved and designed in accordance with the requirements of the utility company.

Harmonics

Non-linear loads have a distorted periodic wave form. Any periodic wave shape can be broken into or analyzed as a fundamental wave and a set of harmonics. Fundamental wave, 60Hz, and harmonics currents exist within the branch circuits and feeders that serve non-linear equipment. The highest magnitudes of currents exist at the 3rd, 5th, 7th, 9th, 11th, and 13th harmonics.

1. The triplet harmonics, including the 3rd, 9th, and 15th, exists only in circuits that have a neutral conductor such as computer circuits and lighting circuits. The triplet harmonics that exist on the secondary side of a Δ-Y transformer will not be transformed to the primary side of the transformer. Transformers with *K* ratings of 4–50 are specifically designed to handle these harmonic currents.
2. Variable frequency drives generate harmonic other than the triplets, the highest of which are the 5th, 7th, 11th, and 13th harmonic currents, on the three-phase circuits that feed them. Drive transformers have been specifically designed for VFD loads.

A harmonic analysis should be performed before installing capacitors on a power distribution system that has a significant non-linear load.

Grounding

Grounding for any electrical distribution system can be divided into two areas: system grounding and equipment grounding. These two areas are kept separate from each other except at the point where the system receives its source of power, namely the service equipment or a separately derived system (Δ-Y transformer or generator). Grounding the neutral conductor at more than one point will cause circulating ground currents that could trip ground fault relays and circuit breakers with ground fault detection and can cause noise in the electronic equipment whose enclosures are required to be solidly grounded to the electrical systems equipment ground.

Pharmaceutical plants will become more dependent on reliable electrical power because of technological advances in pharmaceutical equipment and other loads associated with cGMP areas within plant. A reliable electrical power system is not something that just happens. It requires adequate and knowledgeable planning, engineering, design, installation, testing, and maintenance. A failure to provide any one of these will lesson the reliability of the electrical power system and could leave the plant without electricity. A pharmaceutical plant without electricity is very dark, very quite, and very unsafe.

11

Industrial Molecular Farming

Plant biologists are technology-driven, and many complain about ignorance and hesitative conservatism when they encounter a top-down view of molecular farming in discussions with representatives of the pharmaceutical industry. On the other hand, business people are used to looking at the wider picture. They find essential issues such as downstream processing and product quality addressed insufficiently if at all in molecular farming research. Although knowledge and experience of industrial requirements is accumulating in the green biotechnology field, it is often communicated inadequately. Therefore, the first part of this chapter highlights the requirements and expectations of industrial manufacturers, and the second part considers solutions and answers to the issues that have been raised.

Industrial Production: The Current Situation

Although proteins can be expressed in many heterologous production systems, including bacteria such as *Proteus mirabilis*, fungi such as *Pichia pastoris* and *Aspergillus awamori* and insect cells, the pharmaceutical industry has narrowed down process development to a small number of platform technologies:

1. Mammalian cell suspension cultures are the preferred choice for large-scale recombinant protein production in stirred-tank bioreactors. The most widely used systems are Chinese hamster ovary (CHO) cells and the murine myeloma lines NS0 and SP2/0. In half of the biological license approvals from 1996–2000, CHO

cells were used for the production of monoclonal antibodies and other recombinant glycosylated proteins, including tPA (tissue plasminogen activator) and an IgG_1 fusion with the *tumor necrosis factor* (TNF) receptor, the latter marketed as Enbrel.

2. The bacterium *Escherichia coli* is preferred for the production of small, aglycosylated proteins like Insulin, Proleukin (interleukin-2), Kineret (interleukin-1 receptor antagonist), Neupogen (granulocyte colony stimulating factor, G-CSF) and Interferon-beta.
3. The yeast *Hansenula polymorpha* serves as the production system for a recombinant hepatitis B vaccine and yeast is also used to produce granulocyte-macrophage colony stimulating factor (GM-CSF), marketed as Leukine by Schering AG.

The restriction to a small number of platform production systems is driven by regulatory affairs, risk-benefit evaluation and time constraints that disfavor new production systems. Sauer *et al.* have demonstrated that choosing a current platform system can significantly decrease process development timelines. The situation is different for vaccines and tissue-replacement products, which are produced in many different cell lines using specialized media and cultivation systems. However, it is unlikely that these specialized technologies will be used for bulk production. Consequently, molecular farming has a chance to build up a new platform technology for bulk products if it offers strong advantages, meets requirements and solves current drawbacks in existing production systems.

After the approval of the first product, recombinant insulin, in 1982, progress in the development of new recombinant protein pharmaceuticals was slow. The number of biotechnology-derived drugs and vaccines approved by the US Food and Drug Administration (FDA) has increased significantly only since 1995. More recently, sales of biologics have skyrocketed, e.g. from $900 million in 1999 to an estimated $3.5 billion in 2001 for monoclonal antibodies. The annual global market for biopharmaceuticals is estimated to have increased from 12 billion US$ to 30 billion US$ in 2003. 500 candidate biopharmaceuticals are undergoing clinical evaluation and over one hundred protein-based therapeutics are in the development pipeline for the next few years. Forty of these are likely to reach the market by 2005. Also, many monoclonal antibody products are currently in clinical trials, with the probability of 5–10 new antibodies being approved each year. Das and Morrow (2002) have predicted that sales of recombinant antibodies will reach $8 billion by 2004.

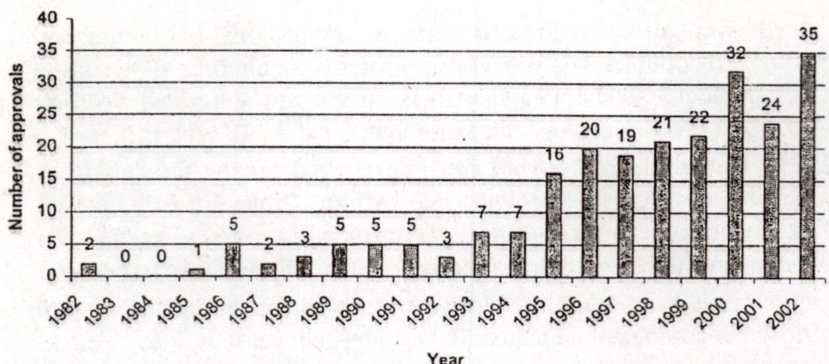

Fig. 11.1. The number of new biotechnological drugs approved each year from 1982-2002.

The major issue facing current production technologies is the need to increase capacities and the related investments sharply. A typical manufacturing facility costs 150–400 million and takes four years to build. Due to regulatory guidelines, equipment and certain materials are dedicated to one specific product. For instance, chromatographic material used for the purification of one protein cannot be used for a different one. The production facility generally needs to be separated from other processes, so only a few companies have multi-purpose equipment. Consequently, it is very difficult to switch from the production of one protein to another, and even more difficult to produce two or more proteins in parallel. This means that the economy of scales plays an important role, i.e. it would be better to produce the most successful protein in the largest amounts. Unfortunately, one cannot predict which of several drugs will be the most successful. In phase III of clinical testing, the decision to build up a production line is difficult to make, since half of the projects fail at this stage. In addition, the time and effort required to obtain approval by the FDA slows down the entire process and demands further resources. As a consequence, many companies hesitate to invest in new production facilities and then suddenly run out of production capacity when therapeutics are ready to be produced in larger volumes.

Manufacturing capacity was estimated to be 575,000 liters in 2002 and was predicted to increase to 1.1–1.4 million liters by the end of 2005. However, demand is expected to increase quicker than production capacity. The urgent bottlenecks are highlighted by two examples: the \$1 million monthly reservation fee Abgenix is paying to the contract manufacturer Lonza according to Arthur D. Little experts and the problems faced by Immunex when the company was unable to meet the increasing demand for Enbrel, a rheumatoid arthritis drug. Immunex share prices plummeted by nearly 75% between August 2000 and August 2001, and the company had to purchase unused capacity at a contract manufacturer from Med-immune.

As a consequence, Immunex decided to invest $400–500 million in November 2001 to build a new facility in Rhode Island and also gained access to a Wyeth plant in Ireland with expected completion in 2005. As similar manufacturing bottlenecks are not unknown to other companies, several have decided to add to their capacities. One might speculate whether low capacities at IDEC might have favored a fusion between IDEC and Biogen.

In addition to the urgent problem of capacity, manufacturers have to cope with the operating costs of production, which are increased by the need for skilled personnel and expensive media components. Another cost driver is the inherent contamination risk when using mammalian cell culture systems. All materials must be checked closely for bacterial and viral contamination, and the presence of prions and endotoxins. This affects not only the manufacturing process, but also downstream materials and even *human serum albumin* (HSA) used for formulations. In the end, production costs add up to $100–1000 per gram of therapeutic protein.

Because of the financial problems faced by public health insurance companies in Europe, there is general pressure to cut the prices of expensive biologics. Political systems support biotechnology start-ups that promise to produce cheaper generic biologics once the patents run out. In Europe, the generics company Sandoz (part of Novartis) has already filed for approval of a generic version of the expensive human growth hormone called Omnitrop (somatropin). The company was able to submit its application after the European Commission adopted a regulatory route for biogenerics in mid-2003. And one can imagine that politicians would similarly pave the way for generic therapeutics produced by Molecular Farming.

Additionally, decreased prices would enable developing countries to afford these drugs. To defend biggest- selling biological drugs against generic versions, the established biotech companies attempt to extend molecule patents by filing patents on production methods. Consequently, there is a great interest in circumventing patents by using new production technologies provided they also offer cost incentives. The expectation is that Molecular Farming might be such a new production technology.

Expectations

As a consequence of the above, interest in molecular farming stems from anticipated time and cost savings once the technology is established. The time aspect has three components. The first time constraint occurs in development. Although the development phase is beyond the scope of this chapter, it should be mentioned briefly that choosing the correct protein to produce is a matter of *time to production.* In the case of antibodies, one would like to produce small amounts of several antibodies, eventually selecting those with the highest stability, affinity and optimal performance according to any other relevant criteria. In the case of vaccines, several protein fragments must be produced to test immunogenic efficacy. Using mammalian cell culture, it generally takes 6–12 months to produce a sufficient amount of recombinant protein for toxicity and other crucial tests. Molecular farming might offer an advantage in this respect, since small but sufficient amounts of purified recombinant protein can be obtained quickly by multiple routes.

One option is to use viral vectors that produce high levels of protein in leaves, albeit transiently. This has been demonstrated in the case of hepatitis B virus surface antigen and single chain antibodies using a tobacco mosaic virus vector. In order to produce a full-size monoclonal antibody, *Nicotiana benthamiana* plants were co-infected with two recombinant tobacco mosaic virus constructs encoding the heavy and light chains, respectively. The overall yield can be higher than with other methods because the viral infection spreads to all cells, resulting in a large number of transgene copies through virus replication. This speedy production platform enabled Large Scale Biology Corp. (LSBC) to develop a personalized vaccine against B-cell *non-Hodgkin's lymphoma* (NHL), a disease with immunologically unique tumors for each patient. The expectation is that each patient can be vaccinated with a plant-expressed anti-NHL scFv manufactured specifically for that individual from the patient's tumor-associated antigens.

Another option, transient transformation, is more familiar to plant biologists. Leaf tissue is infiltrated by *Agrobacterium tumefaciens* carrying a T-DNA vector encoding the protein of interest. The transgene is placed under the control of a plant promoter, which is activated after the T-DNA is transferred into the host plant cell. This method has been shown to be very efficient in the case of tobacco. Agroinfiltration can generate milligram amounts of a recombinant protein within a week. An advantage is that stable plants can be generated in parallel. For plants that can be propagated vegetatively, large amounts of biomass can be produced rapidly for the initial extraction and validation of the heterologous protein. In conclusion, small amounts of the therapeutic protein for pre-clinical tests could be delivered more quickly than is the case with CHO cells, provided that transformation of the host plant species is rapid and efficient and the protein is produced in leaves.

The second time constraint is linked to upscaling. Certainly, scaling up production in plants that produce hundreds of seeds each is one of the greatest advantages of molecular farming. Even for plants propagated by cuttings or tubers (e.g. potato), the agrobiotechnology industry has shown that scale-up is still rapid and economical, since it simply involves increasing the crop acreage and storage capacity. In contrast, at least four years are required to build a new production line for mammalian cell culture and this is a risky venture.

The third time constraint depends on whether the product can be extracted from seeds or fruits. This uncouples protein expression and purification. Large batches of seeds containing the recombinant protein can be produced and stored at low costs. Provided the protein remains stable in the stored seeds, purification can be carried out on demand or shifted according to free capacities. The advantage of one large harvest, with seeds mixed to uniformity, is that this allows production on demand. In contrast, mammalian cell culture is prone to minor batch-to-batch variations in yield and quality. Each fermentation batch has to be processed immediately through down-stream purification. Due to its biologically limited stability storage times had to be minimized.

Probably the most exciting aspect of molecular farming from the industrial perspective is the low initial capital investment compared with mammalian cell culture production. The reduced capital input reflects the lower costs of laboratory equipment and materials for plant molecular biology. For example, the costs involved in establishing small, non-sterile greenhouse facilities are dwarfed by those required for a sealed pilot fermentation plant.

The ultimate decision for big capital investment comes when therapeutic biologics enter phase III of clinical testing. The protein used for phase III testing has to be produced with a process that is identical to that envisaged for future routine production. The process and the equipment cannot easily be changed at a later stage due to regulatory constraints. For instance, it is not possible to switch simply to a fermenter of doubled volume. Alternatively, one could duplicate the process facilities, but this would take years. The supply of a product that has passed phase III might become difficult if the drug sells well and if only minor capacities were implemented during production process design due to financial constraints. However, it is critical to exploit successful products to the maximum extent in an industry in which only 30% of launched drugs return a positive *net present value* (NPV). This above-average income offsets other products in company portfolios that never provide a return on capital employed (ROCE). The above-mentioned case of supply shortage for Enbrel is such an example. It is evident that molecular farming allows managers in pharmaceutical companies to scale up production more easily if sales are higher than expected. Doubling production is just a matter of sowing or planting twice as many plants.

The production of edible vaccines in fruits and vegetables provides another large cost saving. Given the pre-requisite of guaranteed immunogenicity and adequate vaccine level in the fruit, the expensive purification stage could be omitted and would reduce costs considerably. In the case of β-glucuronidase produced in corn, 88% of operating cost has been attributed to protein extraction and downstream processing. A big share which could be saved in case of edible vaccines.

Improved safety is often highlighted by researchers as one of the advantages of molecular farming. Indeed, contamination of mammalian cell cultures remains a significant issue since antibiotics are used in a number of large-scale industrial processes. However, microbial contamination is also an issue in the case of plants. This concern was already addressed for example in regulatory guidelines and is a subject of current research in the area of medicinal plants. The food industry has already developed some cost-effective solutions. For example, lye peelers, which use 5–15 % sodium hydroxide solutions and elevated temperatures, effectively remove potato skin and would inactivate soil-borne pathogens on the tuber surface.

In addition, the regulatory authorities are concerned about contamination with viruses and prions, such as the causative agent of *bovine spongiform encephalopathy* (BSE), which could be present in mammalian cell cultures. It is necessary to check all material used for protein production and purification to eliminate such contaminants. The regulatory authorities have recommended the use of serum-free, animal protein-free media as well as concomitant changes in cell line development and process design, but a small number of processes still require calf serum. Instead of *human serum albumin* (HSA) purified from human blood, recombinant HSA can be used. Furthermore, one has to show that chromatographic filtration steps reduce viral titers by several orders of magnitude. It is argued that molecular farming would not be subject to such controls because plant viruses have been eaten by humans since prehistoric times with no adverse effects, excluding the existence of potentially harmful pathogens. However, draft guidelines from the EMEA show that concerns about viral contamination persist although not scientifically justified. The final outline of the regulatory guidelines remains to be seen.

In summary, alternative production systems are attractive if they provide solutions for the above-mentioned problems. Molecular farming is one such system, with compelling advantages in terms of time, money and safety. In addition, however, it must fulfill the requirements discussed in the following section.

Requirements

Consistent efficacy and biological equivalence are the conditions *sine qua non* for regulatory authorities and consequently they are the major concern of executives in the pharmaceutical industry. Miele, an FDA official, stated in 1997 that "Recombinant macromolecules produced in plants should be biochemically, pharmacologically and clinically comparable to their counterparts produced in traditional cell substrates or animals or purified from human sources, if such products are available." Unfortunately, this paper was largely ignored by most molecular farming researches, and very few studies have been carried out to assess the biological equivalence of proteins produced in plants and mammalian cells. A belief that the approval of plant-derived pharmaceuticals is outside of the scope of research might jeopardize the whole field of molecular farming. Managers will turn down any molecular farming project where there is a high risk that the product will not enter the market. Answering the following key issues will minimize the risk of failure.

Equivalence of the Recombinant Product to the Original Protein

A panoply of physical, chemical and biological or immunological tests should be instituted for the purpose of establishing comparability between products derived from plants and those from animals or humans. A combination of SDS-PAGE and mass spectrometry to determine the molecular weight, protein sequencing, isoelectric focusing (IEF), H PLC, peptide mapping (e.g. tryptic mapping) and carbohydrate mapping should be applied. *In vitro* potency assays can often be used to screen for possible differences in bioactivity. In relevant animal models, the equivalence of plant-derived products can be shown in terms of pharmacokinetic studies and biological activity. In addition to equivalence, batch-to-batch consistency needs to be shown. The recent case of Eprex versus Procrit (both erythropoeitin) had alarmed authorities as serious side effects (red cell aplasia) occurred in patients treated with Procrit made at a J&J factory in Puerto Rico while none occurred with batches made at Amgen's facility. This has lead to speculations that differences in the manufacturing process had altered the final product.

Processing in the Endoplasmic Reticulum (ER)

Many proteins have to pass through the ER and Golgi apparatus to be processed correctly, to fold efficiently or to form disulfide bridges. These proteins need be expressed as pre-proteins, which contain an N-terminal signal peptide that must be cleaved off during translocation into the ER. The efficacy of plant signal peptides in targeting fusion proteins to the ER has been demonstrated by several methods, e.g. electron microscopy. The proper cleavage of signal peptides from heterologous pre-proteins has been reviewed. Most recombinant proteins are cleaved correctly, thus preventing the accumulation of pre-proteins instead of mature proteins, although some exceptions have been reported. Interestingly, expression of the entire prepro-HSA protein in tobacco and potato resulted only in partial processing of the precursor, and the secretion of pro-HSA into the medium. In contrast, fusion of the tobacco PR-S protein signal sequence directly to the HSA gene resulted in the production of mature HSA, indistinguishable from the original human protein. The tobacco Pr1 b signal peptide was fused to the cholera toxin B subunit, and correct processing was demonstrated by SDS-PAGE. Furthermore, the antigenicity of the CTB protein was identical to its native counterpart as judged by immunodiffusion and immunoelectrophoresis. Only in a few cases has proper cleavage of the plant signal peptide been demonstrated by N-terminal amino acid sequencing: (i) the proteinase inhibitor II signal sequence fused to HSA produced in potato, (ii) the rice α-amylase signal peptide attached to a single-chain antibody fragment expressed in *N. benthamiana*, (iii) the barley α-amylase signal peptide attached to avidin expressed in maize, and (iv) the rice RAmy3D signal peptide attached to human lysozyme expressed in rice cell culture. In this last report, the authors demonstrated that the recombinant enzyme had the same molecular mass and isoelectric point as native human lysozyme. The integrity of purified aprotinin, commercially produced in transgenic maize seeds, was thoroughly analyzed and compared to native bovine aprotinin with respect to molecular weight, the presence of disulfide bonds, pI, N-terminal amino acid sequence and trypsin inhibition activity. The proper formation of disulfide bridges was shown for recombinant trout growth hormone expressed as a fusion with the PR1 b signal peptide in tobacco.

Some reports indicate that ER-targeting, using a C-terminal KDEL sequence, is necessary to achieve high protein levels. It has been suggested that proteins accumulating in the cytosol suffer proteolytic degradation and therefore do not reach very high levels. Interestingly, adding a KDEL signal to several scFv antibodies resulted in considerably improved expression levels (0.2%) even though the antibodies remained exclusively in the cytosol. This indicates that the KDEL signal may generally stabilize proteins. Even so, the level of scFv accumulation with ER-targeting and the KDEL signal was fivefold higher (1%). Therefore, many research teams have used the C-terminal addition of KDEL to further improve the yield of ER-targeted proteins. Some exceptions of the rule are known. For instance, the recombinant heat-labile toxin of enterotoxigenic *E. coli* joined to a barley α-amylase signal sequence yielded higher protein levels if targeted to the maize cell surface than if retained in the ER. Although many researchers favor the addition of the KDEL or KDEI retention signal, most pharmaceutical companies would need the effect of the KDEL tag on immunogenicity to be assessed thoroughly. For instance, an ER-retained recombinant hepatitis B virus surface antigen (HBsAg) containing a KDEL tag was favored as a vaccine candidate because mice immunized with this protein showed a stronger immune response than those immunized with unmodified plant-derived HBsAg. However, the immunoreactivity of mice against uncleaved signal peptides was not assessed. The fact that a recombinant protein with a KDEL-tag is not equivalent in the strict sense to the native protein is often overlooked. One should be aware that it is necessary to clarify whether the addition of KDEL might influence biological activity when used in therapy. Most pharmaceutical companies would prefer to have smaller amounts of the correct protein.

Glycosylation in the Golgi

Targeting recombinant proteins to the secretory pathway is essential for correct folding, the formation of disulfide bridges and glycosylation. Glycosylation influences many properties of recombinant proteins including biological activity, folding, solubility, stability and blood clearance. Glycosylation of the Fab regions has been shown to affect the binding activity of some monoclonal antibodies, in contrast to glycosylation in the Fc region, which rarely has an impact on immunoreactivity. However, the correct glycosylation of the C_H2 domain is necessary for the efficient binding of IgGs to the Fc receptor of immune cells. Therefore, in general, recombinant proteins produced for pharmaceutical purposes should have the same glycosylation profile as the native human protein.

CHO cells have been favored because the glycans they attach to proteins are very similar to those of human cells. If other platform technologies result in differing glycosylation patterns, new regulatory approval for the recombinant proteins must be obtained, certainly in the case of generic biologics. However, one has to bear in mind that glycosylation patterns vary in any heterologous system. The glycosylation of tPA was shown to depend on CHO cell growth rate and temperature. Similarly, growth conditions have been shown to influence the glycosylation of an IgG antibody expressed in tobacco leaves. Interestingly, the galactosylation of serum IgG glycans increases also in healthy pregnant women. Reportedly, MedImmune's Synagis, a humanized monoclonal antibody, consists of several glycoforms. This indicates that a certain degree of microheterogeneity would be acceptable by the FDA.

While the above reflect rather minor variations, the major differences in glycosylation between plants and animals might not be seen in the same light. Indeed, plants (like insects and yeast) do not incorporate sialic acid into their glycan chains, and the fucose linkage is different to that found in mammals. Furthermore, plant glycans contain immunogenic xylose residues, which are never found in human glycoproteins. For instance, recombinant avidin obtained from transgenic maize was glycosylated at Asn-17 with a carbohydrate chain smaller than that of chicken egg avidin. Similarly, the N-linked oligosaccharides attached to erythropoeitin are smaller in tobacco than in mammals. The N-glycan composition of a murine monoclonal IgG_1 antibody was studied detailing and compared with the glycan structures of the same monoclonal antibody expressed in tobacco. Although the N-glycosylation sites were the same in both cases, different glycans were detected, and the tobacco glycans contained terminal xylose and fucose residues. The unusual sugars could be detected easily using the lectin concanavalin A or anti-β(1,2)-xylose or anti-α(1,3)-fucose antibodies. Notably, the analysis of another IgG antibody produced in tobacco revealed that a proportion lacked terminal xylose and fucose residues in contrast to endogenous proteins. An ER-retained anti-HBsAg antibody had predominantly oligomannoside type N-glycans attached to the same N-glycosylation sites as seen in the mouse antibody. However, complex type N-glycans were also detected, containing α(1,3)-fucose attached to the core GlcNAc residue. Due to the presence of non-mammalian xylose/fucose residues, plant-derived drugs might be recognized as foreign antigens in mammals.

Fortunately, this seems not to be the case: no immunogenic reaction was seen in mice when they were immunized subcutaneously with the Guy's 13 antibody expressed in tobacco. Furthermore, in human clinical trials, a modified version of the tobacco-derived Guy's 13 antibody was administered on six occasions, and did not elicit an immunological reaction against the plant-specific N-linked glycans. Similarly, no adverse reactions were reported when a soybean-derived antibody directed against herpes simplex virus 2 was delivered to the mouse vagina. Similar results were also reported by EPIcyte: a mouse monoclonal antibody produced in maize did not elicit an immune response in mice. In the few clinical trails conducted so far problems with allergic reactions towards plant-derived drugs have not been reported. It was suggested that plant glycans are known to the immune system as components of our daily food intake. The rationale is that food ingredients cross the intestine barrier into the blood stream, and therefore cannot stimulate a constant immune response. More recent results oppose this view: antibodies specific for core-fucose and core-xylose-epitopes were found in the blood of non-allergic donors, probably resulting from food or environmental sensitization.

Differential Glycosylation – Implications on Immunogenicity of vaccines

Whereas immunogenic reactions against the therapeutic protein should be avoided in the case of plant-derived antibodies, an immune reaction is needed for vaccines expressed in plants. However, the modified glycosylation of plant-derived vaccines compared to the original protein might change the immunogenicity of the vaccine and might even cause allergies. Some indications for altered immunogenicity were found when horseradish peroxidase was injected into mice and rats. They raised not only antibodies against the protein moeity, but also against core-fucose and core-xylose-epitopes.

On the other hand, for the hepatitis B surface antigen, the glycan was shown not to be required for immunogenicity, and in contrast to the soybean-derived vaccine the yeast-derived vaccine has no glycosylation. Interestingly, the glycan chains are lost with storage of the partially purified extract from vaccine-producing soybean cell cultures. Furthermore, engineered mutations that increased the amount of N-linked glycosylation on the coat protein of human T-lymphotropic virus (HTLV-I) did not alter the immunogenicity of the virus, but enhanced immune recognition.

Previous studies had shown that the measles virus hemagglutinin (H) protein is highly susceptible to altered glycosylation. Glycosylation is required at two or more of the four sites usually glycosylated in the native protein, otherwise protein folding, stability and protease susceptibility are adversely affected. The H protein produced in transgenic tobacco and carrot had a molecular weight lower than the native protein, probably due to differences in the glycosylation pattern, but it retained its antigenic and immunogenic properties. This indicates that plant-type glycan structures did not affect the protein's conformation. This is further corroborated by the recognition of the plant-derived protein using monoclonal antibodies or human serum antibodies produced in response to a wild-type measles infection. Oral as well as intraperitoneal immunization of mice induced high titers of IgG antibodies that neutralized the virus in vitro. Interestingly, antibodies induced by injecting transgenic carrot leaves extract were of both the IgG_1 and IgG_2 a subclasses, whereas the immune response in mice injected with measles virus hemagglutinin protein produced in mammalian cells was essentially restricted to IgG_1. However, overall antibody levels were comparable. Taking all the data together, it seems that differences in glycosylation have little impact on the immunogenicity of plant-derived vaccines. Their efficacy has also been demonstrated in clinical tests. Nevertheless, the pharmaceutical industry would prefer unchanged glycosylation patterns in order to avoid possible side effects caused by non-human glycan structures.

Glycosylation and Stability

Glycosylation affects protein stability as well as immunogenicity. A prominent example is *tobacco-derived erythropoeitin* (EPO), which possesses smaller N-linked oligosaccharides compared to commercial EPO and lacks sialic acid residues. The plant-derived protein was shown to be more active in vitro i.e. inducing the differentiation and proliferation of erythroid cells stronger than commercial EPO. Strikingly, this difference was eliminated if the commercial EPO was desialylated. However, neither the tobacco-derived protein nor the desialylated recombinant EPO showed activity in vivo. The authors suggested that EPO with glycans lacking terminal sialic acids is probably trapped by the asialoglycoprotein receptor on hepatocytes and excluded from the circulation before it can exert biological activity in rats. Intriguingly, the contrary effect i.e. prolonging half-life of an EPO analog was achieved by increasing sialic acid content. This modification was considered as an advantage by the FDA.

Glycosylation also affects the stability of antibodies expressed in plants. For example, an ER-retained antiviral monoclonal antibody was barely detectable 10 days after injection into mice, whereas the same antibody obtained from murine/human hybridoma cell lines was still abundant in blood samples. This finding was surprising since there was no difference between the antibodies in terms of their rabies virus neutralizing activity, and no differences in efficacy measured in terms of post-exposure prophylaxis for hamsters injected with rabies virus. The higher stability of the mammalian antibody might again be dependent on terminal sialic acids in complex N-glycans. In contrast to mammalian antibodies, the ER-retained anti-viral antibody from tobacco displayed predominantly oligomannose-type N-glycans (90%), although no $\alpha(1,3)$-fucose residues were detected. Oligomannose-type N-glycans are likely to be recognized by mannose receptors on e.g. liver macrophages, resulting in increased blood clearance of the plant-derived antibody. However, the shorter half- life of the tobacco-derived antibody may offer also an advantage for therapeutic use because there will be less interference between passive and active immunity compared to the current commercial antibody vaccine. Studying the degradation of antibodies expressed in plants, Stevens et al. discovered that the tobacco-derived antibody was less stable in plant extracts than the control antibody produced by hybridoma cells. The examples given emphasize the importance of glycosylation to ensure stability of plant-derived biological drugs.

For other plant-derived antibodies, stability was shown to be similar to mammalian counterparts. For instance, a humanized anti-herpes simplex virus monoclonal antibody (IgG_1) was expressed in soybean and showed stability in human semen and cervical mucus over 24 h similar to the antibody obtained from mammalian cell culture. In addition, the plant-derived and mammalian antibodies were tested in a standard neutralization assay with no apparent differences in their ability to neutralize HSV-2. As glycans may play a role in immune exclusion mechanisms in mucus, the diffusion of these monoclonal antibodies in human cerival mucus was tested. No differences were found in terms of the prevention of vaginal HSV-2 transmission in a mouse model, i.e. the plant-derived antibody provided efficient protection against a vaginal inoculum of HSV-2. This shows that glycosylation differences do not necessarily affect efficacy.

Some research groups prefer IgAs because they are more stable than IgGs and more resistant to proteolysis. Plants can assemble secretory IgAs, which consist of four chains (heavy and light chains, J chain and secretory component). Additionally, secretory IgAs are more efficient at binding antigens. These characteristics allow IgAs to treat mucosal sites such as the gastrointestinal tract against infections. As repeated large doses of antibody are required for topical passive immunotherapy, transgenic plants could be the only cost-effective production system.

In conclusion, glycosylation and processing should result in plant-derived pharmaceutical proteins mainly identical to wild-type protein even though different forms may have exactly the same characteristics as the original protein. However, differences in glycosylation might be acceptable by the FDA if the isoforms offers advantages e.g. Aranesp (darbepoetin alpha) and glucocerebrosidase.

Equivalence of Enzymes

Drugs used in replacement therapies are often enzymes, as exemplified by glucocerebrosidase and iduronidase. The treatment of Gaucher patients is based on the external supply of the recombinant enzyme glucocerebrosidase and costs $100,000–400,000 per annum despite the recent approval of CHO-derived cerebrosidase. Enzymatically active glucocerebrosidase has also been obtained from transgenic tobacco and purified using an epitope tag. The plant-derived enzyme is comparable to the human or CHO-derived glucocerebrosidase based on kinetic studies and affinity binding characteristics. Glucocerebrosidase from tobacco comigrates with the human placental-derived enzyme in SDS-PAGE, and is glycosylated. Other examples for equivalence of plant-derived enzymes are gastric lipase and neomycin phosphotransferase II (NPTII). The equivalence of latter enzyme produced in *E. coli*, cotton seed, potato tuber and tomato fruit has also been assessed. Microbial and plant-produced NPTII proteins have comparable molecular weights, immunoreactivity, epitope structures, N-terminal amino acid sequences and biological activity. Interestingly, all NPTII proteins remained unglycosylated irrespectively of the production organism.

Degradation

Reliable extraction protocols are needed to prevent the degradation of plant-derived antibodies. Many researchers use protease inhibitors, such as phenylmethanesulfonylfluoride (PMSF) leupeptin or mixtures thereof. Following detailed analysis, however, the degradation appears to occur in planta rather than during extraction. This hypothesis is corroborated by the finding that antibody degradation fragment patterns are not changed following the addition of protease inhibitors or protein protective agents. Some researchers have noted a protein band of 120–125 kD often seen in preparative SDS-PAGE gels for plant-derived antibodies. Van Engelen *et al.* and Stevens *et al.* interpreted the 125 kD fragment as $F(ab')_2$-like fragment whereas Sharp and Doran found cross-reaction of this band with an anti-Fc antibody. But all authors agree that this fragment and others stem from proteolytic activity inside the plant cells, most probably starting with the proteolytic removal of (part of) the constant region (Fc). Interestingly, this break-down results in relatively stable products. In general, it seems that degradation is lower for ER-retained proteins in storage organs. For example, an ER-retained scFv purified from potato tubers in total protein extract was found to be stable for at least 16 days at 4°C. In summary, the role of degradation needs to be studied more extensively for plantibodies. Avoiding degradation might also improve yields in a more efficient way than simply increasing expression.

For orally delivered plant vaccines some crop processing might be necessary to ensure consistent antigen dosage. The subsequent loss in compartmentalization could expose the antigen to proteases, polyphenol oxidases and plant phenolics. Consequently, immunogenic epitopes could be destroyed or changed. Serum-derived hepatitis B surface antigen (HBsAg) proved to be remarkably protease resistant, which was attributed to its extensive disulfide cross-linking yielding dimers and higher multimers. Unfortunately, this extensive cross-linking does not occur in the plant-derived HBsAg resulting in degradation of the HBsAg dimers in plant extracts. Different proteinase inhibitors, including leupeptin, aprotinin, E-64, pefabloc and pepstatin had no effect on HBsAg stability in potato extracts. Only the combination of leupeptin and β-mercaptoethanol protected the antigen. Interestingly, under optimized detergent conditions, protein stability was extended to 1 month in tomato, but not potato extracts. The N-terminal addition of a signal peptide resulted in a more stable, but uncleaved HBsAg fusion protein that formed multimers and was more potent than unmodified plant-derived HBsAg. Despite the protease-sensitive nature of potato-derived HBsAg, uncooked transgenic potatoes elicited an increased immune response when compared with a similar oral dose of commercial yeast-derived rHBsAg. Similarly, rHBsAg partially purified from transgenic tobacco leaves generated a qualitatively similar immune response in mice when compared to yeast-derived rHBsAg. This indicates that despite some degradation, both the B- and T-cell epitopes of HBsAg are preserved when the antigen is expressed in transgenic plants. Other stability studies were carried out on clover plants expressing the *Mannheimia haemolytica* A1 leukotoxin 50 fusion protein. After harvest, clover plants were allowed to dry at room temperature for 1–4 days. No degradation of the fusion protein was observed and the protein induced an immune response in injected rabbits.

Efficacy in Clinical Trials

As highlighted by the erythropoeitin example discussed above, efficacy and bioequivalence shown *in vitro* do not guarantee efficacy in vivo. But, the latter is a key requirement for the success of protein therapeutics expressed in plants. One of the concerns regarding the use of plant-derived edible vaccines is that humans ingesting transgenic plants might not respond to immunization because the same plant species is part of their regular diet. To date, a number of clinical trials using edible vaccines have been carried out or are under way. The first human trial was carried out in 1997. Fourteen healthy adult volunteers were given either wild type or transgenic potato, the latter containing a recombinant B subunit of the heat-labile enterotoxin of *E. coli* (Lt-B) as antigen. The raw potatoes were generally well tolerated. Antibody- secreting B cells were detected seven days after eating one dose of transgenic potatoes. Ten of eleven volunteers developed IgGs against Lt-B protein, but neutralizing titers were only achieved in eight volunteers. Similar efficacy was shown in mice ingesting an ER-retained form of Lt-B, although the expression level was quite low (<0.01% of TSP). The mice developed both serum and gut mucosal antibodies specific for Lt-B. In different sets, mice were shown to be partially protected against challenge with the holotoxin after oral delivery of Lt-B-producing potato or maize expressing a codon-optimized Lt-B. In contrast, Lauterslager *et al.* found that oral immunization elicited systemic and local IgA responses only in parenterally primed, but not naive mice.

Many studies have been carried out on plant-derived Hepatitis B surface antigen (HBsAg) showing its equivalence to the commercially available HBsAg from yeast. HBsAg was successfully expressed in tobacco, potato, tomato, soybean, lupin and lettuce, but only transgenic lettuce was used for two human trials with five and twelve volunteers, respectively. After two or three immunizations with the transgenic lettue, all test participants showed low levels of HBsAg-specific antibodies. Unfortunately, protective levels were only transiently reached in two volunteers, and this might have been due to the low expression level of the recombinant HBsAg (<0.01%). No apparent side effects occurred in volunteers within 20 weeks after first ingestion of transgenic lettuce. In pre-clinical animal studies, uncooked transgenic potatoes containing as little as 0.0001% HBsAg had to be administered together with an adjuvant (cholera toxin) due to the limited amount mice could ingest in 24 h. Subsequent parenteral boosting was necessary to reach protective antibody levels. In conclusion, the efficacy of the commercial yeast-derived vaccine (also containing an adjuvant) has not yet been matched by molecular farming in plants.

The *Norwalk virus capsid protein* (NVCP) has been expressed in transgenic tobacco and potato at levels of 0.02–0.07%, but in potato only 50% of NVCP correctly assembled in *virus-like particles* (VLP). Having ingested 2–3 doses of raw transgenic potato, 19 out of 20 volunteers showed a significant increase in the number of specific IgA antibody-secreting cells. But only four and six people developed serum anti-NCVP IgGs and IgMs, respectively. None of the volunteers showed changes in serum anti-patatin IgG after ingestion of transgenic potatoes, which contain patatin as major storage protein. In 2002, another clinical trial was carried out using transgenic spinach expressing epitopes from the rabies virus glycoprotein and nucleoprotein fused to the coat protein of alfalfa mosaic virus. After oral immunization, five of nine naive volunteers and three of five volunteers primed with the conventional vaccine showed significant elevation in rabies-specific antibodies. As it is risky to assess the protective level against rabies in humans, a mouse model was used to test the plant-derived rabies vaccine. After parenteral immunization, mice were protected against challenge infection. As far as animal edible vaccines are concerned, Prodigene Inc. were able to demonstrate the efficacy of a corn-derived vaccine against challenge with swine-transmissible gastroenteritis coronavirus (TGEV). Protection against challenge was also reported for a potato-derived rabbit hemorrhagic disease virus vaccine in rabbits and for a potato-derived VP1-vaccine protecting against food and mouth disease virus. In summary, most studies indicate that efficacy could be still improved, probably through higher expression levels in plants.

In contrast to edible vaccines, plant-derived antibodies and therapeutic proteins are mainly prepared for clinical testing by biotech companies, and fewer details have been published. According to the company's press release, EPIcyte is slated to become the first company to enter Phase I clinical trials with a human herpes antibody (called HX8) produced in plants. Trials in 2003 should show the efficacy of HX8 to prevent the transmission of herpes simplex virus 1 and 2. A phase II SBIR grant has been awarded to EPIcyte, which is allied with Dow. The efficacy of such an antibody from soybean for prevention of vaginal HSV-2 infection in mice has been shown. Tobacco-derived secretory IgAs, marketed as CaroRx, were designed to prevent oral bacterial infections contributing to dental carries. In a clinical trial, the plant-derived IgA gave specific protection against colonization by oral streptococci for over four months. No adverse effects or immunological response against the IgA have been observed in more than 40 patients receiving topical oral application of the IgA. According to another company's homepage, Planet Biotechnology is currently undergoing Phase II US clinical trials under a US FDA-approved Investigational New drug (IND) application. The companies NeoRx and Monsanto Protein Technologies joined to produce a humanized antibody in corn to be used for Avicidine cancer treatment. Unfortunately, all that has been reported about the clinical trails is that the maize-derived antibody behaved similarly to the murine antibody. Some months later, the project was discontinued by the partner Janssen because of a high incidence of severe side-effects in Phase II trials with the murine antibody. No serious adverse side-effects were reported by Large Scale Biology Corp. (LSBC) after completion of phase I clinical trials with their personalized cancer vaccine, by August 2002. Each patient in the study received a tobacco-derived scFv manufactured specifically for that individual from the patient's tumor-associated antigens (non-Hodgkin's lymphoma). 15 of the 16 individuals showed no immune reaction to the plant-derived single-chain antibody and ten mounted a humoral or cellular response to the treatment. Researchers at LSBC are working together with the FDA to design PhaseII/III trials. Even more promising is LSBC's human α-galactosidase for which the FDA granted orphan drug status in January 2003. The enzyme is also produced in tobacco infected with a re-engineered virus and showed positive results in pre-clinical studies using an animal model of Fabry disease. Positive responses to a treatment with corn-derived gastric lipase given to 15 cystic fibrosis patients resulted from a phase II multicenter trial within 2002. Meristem Therapeutics and its exclusive partner Solvay S.A. reported about reduction of fecal lipid content during these clinical trails. According to Giddings *et al.*, Ventria Bioscience started trials with α-1-antitrypsin from transgenic rice and hopes to get product approval by 2004. It will be interesting to see in years to come, if plant-derived antibodies and enzymes will show full efficacy.

Optimal Production System

Stability and degradation are also issues when it comes to choosing the production system. Industry would ask for a plant production system that offers the possibility of storage and transport from the field to the downstream processing facilities. As discussed above, this can be difficult in the case of leaves and leaf extracts due to protein degradation. Losses of enzymatic activity in leaf extracts were observed in a feasibility study. There is one exceptional report concerning an ER-retained scFv antibody expressed in tobacco leaves that was extracted after drying and storage for one week, and was still active. Khoudi *et al.* reported antibody stability in dried alfalfa hay as well as in extracts made in pure water. Seeds as natural storage organs seem to be the better alternative for several reasons: (i) Seeds contain a less complex mixture of proteins and lipids. This is an advantage for purification. (ii) The stability of recombinant proteins is likely to be higher in seeds. Seeds have low protease activity and contain fewer phenolics than leaves. The advanced state of dehydration confers enhanced stability, allowing seeds to be stored for periods of several years without any notable degradation of proteins or loss of activity, as shown for phytase. In rapeseed, a β-glucuronidase-oleosin fusion protein was stable for long periods. In *Arabidopsis*, an scFv antibody expressed under the control of the β-phaseolin and arcelin-5 promoters accumulated to 36% TSP in homozygous seeds and was stable for at least one year, while a level of 2% TSP was achieved using same constructs in *Phaseolus acutifolius*. In tobacco seeds, an active scFv accumulated to 0.7% TSP, and the seeds could be stored for one year at room temperature without protein degradation or loss of antigen-binding activity. Similar stability was observed for another scFv expressed in dry wheat and rice grains. Furthermore, *E. coli* Lt-B and the S antigen of TGEV were stable for longer than a year when expressed in maize seeds and extracted in the germ meal fraction. Avidin was stable in whole maize kernels, but not in flaked material, and could withstand temperatures up to 50°C for at least 7 days without loss of activity. The potato tuber is another natural storage organ, but unfortunately it only contains 2% protein. Artsaenko *et al.* reported a 50% loss of functional antibody after 18 months in storage, starting from a pre-harvest yield of 2% TSP in potato tubers. To fulfill the industrial standard of current logistics in medicinal plant drug production, germplasm with low protease activity is needed to ensure stability for at least one year in storage. Despite the advantage of seed-based production, however, it might not be possible to express all proteins in seeds.

Stable expression of the recombinant protein is another prerequisite. For production in CHO cell culture, narrow specifications of expression level are requested by the FDA. Consequently, plant-derived proteins need to be produced according to a defined scheme and within defined specifications in order to generate material suitable for clinical trials. Therefore, the expression level must be maintained within a defined range. For example, Prodigene Inc. assessed whether the *E. coli* Lt-B protein was uniformly distributed throughout the defatted maize germ by analyzing samples taken randomly from the bulk material. The amount of an IgG expressed as percentage of total soluble protein was analyzed in tobacco leaves from the top (0.15–0.24%), middle (0.13–0.19%) and base (0.14–0.21%) of the plants, indicating no dependence on leaf age. However, it will be always difficult to control expression levels in the field. Therefore, expression in a controlled environment is more appropriate to ensure specified expression levels. Post-harvest induced expression would fit to this requirement.

The ideal crop should be amenable to transformation and regeneration, and should also facilitate rapid scaling up of production when required. Tobacco and rapeseed are prolific seed producers (up to one million seeds produced per plant in the case of tobacco). However, it would be necessary to establish a homogenous line after several generations, since even in the third generation, large differences between progenitor lines can occur and selection and back-crossing with elite germplasm over several generations might be necessary to achieve economical protein yields. Consequently, seed propagation makes it necessary to monitor the stability of transgene expression after re-amplification from the master seed bank. Therefore, in terms of speed and fidelity, crops reproduced by vegetative propagation (e.g. cuttings or sprouts) might be preferred. Additionally, the guideline draft by EMEA favors vegetative propagation. For example, no apparent loss in the ability to synthesize and accumulate a fully functional recombinant antibody was reported in alfalfa plants produced through stem propagation.

Another valuable feature of the ideal production system would be a high biomass yield in combination with established, efficient harvesting and processing technologies and infrastructure. Recovery of the therapeutic protein often involves grinding the plant tissue and immediate cold buffer extraction and removal of solids. The availability of harvest and recovery technologies (e. g. milling or oil-extraction) would reduce costs and investments at the beginning. A crop would offer a considerable benefit if the recombinant protein was expressed at high levels in a defined part of the plant, e.g. the wheat germ or oil-fraction, which can be easily obtained using established technologies. For example, the fractionation of transgenic seeds by dry and wet-milling removed seed components (fiber, starch, oil) that did not contain the recombinant protein.

The maize embryo fraction is rich in soluble protein and can easily be separated from other seed tissue to increase the concentration of the recombinant protein. This enriched fraction would decrease storage volumes, transport costs and the extract volume to be handled in the subsequent steps. Additionally, fractionation would decrease the complexity of the extract and reduce the level of plant metabolites that interfere with downstream processing thus facilitating purification. Furthermore, fractionation procedures and milling will result in a defined particle size distribution which is helpful to design first filtration or cleaning steps prior to protein purification. For established crops, a large selection of pesticides is available that would facilitate the selection of chemicals appropriate for the production of plant pharmaceuticals. Finally, the ability to grow the crop of choice on both hemispheres could ensure constant supply and reduce storage capacities and the risk of supply problems due to harvest losses in one region.

Importantly, the optimal production crop must offer a high standard of safety, also in view of liabilities. Since biosafety is discussed in more detail in chapter 16, I will only briefly summarize the expectations of pharmaceutical industry:

(i) *Transgene containment*: Out-crossing to wild relatives or non-transgenic plots of the same crop should be prevented by appropriate choice of the crop plant and the location of the transgenic plot, by physical containment, or by genetic containment mechanisms.

(ii) *Identity preservation*: A crop with a visible marker, such as a defined seed color, would help to prevent mixing with consignments of the same crop during harvesting, transport and processing.

Examples could include white tomatoes, black barley, red carrots and pink potatoes.

(iii) *Product containment*: A plant variety that is non-toxic, but has an unpalatable taste, would help to prevent involuntary ingestion of the recombinant protein. Strategies that direct expression to inedible organs or causes the protein to be expressed after harvest would provide a similar measure of safety.

Post-harvest Expression

Post-harvesting technologies have already been established. CropTech Corp. is using the inducible MeGA promoter to produce glucocerebrosidase in tobacco leaf tissues. Although expression levels vary depending on the plant line and the precise induction protocol, over 1 mg of the enzyme per gram fresh weight has been achieved in crude extracts. Malting of rice or barley seeds is a technical process analogous to natural seed germination. Under defined malting conditions, many parameters like protein content and protein quality can be controlled. The companies Maltagen and Ventria Bioscience took advantage of the classical malting technology and developed protein production systems using transgenic grains. Under the control of germination-specific promoters, coordinated expression of several transgenes (e. g. light and heavy chain genes to produce a full size antibody) can be achieved. The germination-specific promoters allow exact regulation of the expression level during malting. The influence of the environment on the integrity, stability, glycosylation and expression level of the heterologous protein is excluded since the promoters (e.g. α-amylase promoter) are inactive during the growth period in the field, and active only under controlled malting conditions. One expects that this would avoid transgene silencing problems especially arising when (multiple copies of) strong promoters are used to achieve high-level expression.

Purification

In contrast to the protein recovery methods discussed above, protein purification is still based predominantly on laboratory-developed procedures that are often not directly scalable because of the high costs of the chemicals employed, the difficulties in controlling foaming, and pumping problems related to non-homogeneity and viscosity. For example, Fischer *et al.* developed a three-step protocol for the recovery of antibodies from extracts of tobacco suspension cells, starting with cross- flow filtration followed by protein A affinity chromatography and gel filtration. More than 80% of the expressed recombinant antibody was recovered. Some studies have been carried out concerning the choice of defatting solvents and the recovery of recombinant proteins from canola by cation exchange chromatography. However, highly efficient purification protocols for molecular farming are still lacking, even though such protocols are absolutely necessary with regard to economic and regulatory affairs. Protein recovery and purification costs are estimated to account for 88% of operating costs. Similarly, for the commercial production of insulin in *E. coli*, chromatography accounts for 30% of operating expenses and 70% of equipment costs. Therefore, the low production costs of molecular farming would not counterbalance the higher downstream processing costs associated with plant tissue if the efficiency of purification from plants could not be increased dramatically.

Affinity tags such as His_6, Myc and FLAG have been used to facilitate purification by affinity chromatography. However, therapeutic proteins containing additional affinity tags will certainly not obtain approval. Therefore, economical and viable methods to cleave off the affinity tag are required. A slightly different, and very elegant approach is the use of an oleosin tag in combination with simple flotation-centrifugation technology. SemBioSys Inc. has developed an oleosin-fusion platform that includes cleavage of the oleosin-fusion protein with proteases like factor Xa in a cost-efficient manner.

An intriguing problem in recombinant protein purification is the presence of co- purifying proteins. An excellent study has been carried out on the removal of *corn trypsin inhibitor* (CTI) which co-purified with aprotinin on a trypsin-agarose affinity column. After grinding and milling, protein extraction at pH 3 reduced the amount of CTI in the extract and increased the aprotinin content in the mass fraction. After subsequent filtration and affinity adsorption, the remaining CTI was captured on an agarose-IDA-Cu^{2+} column while the recombinant aprotinin was collected in the flow-through with a purity of at least 79%.

The most important issue in downstream processing is often neglected: quality management. Similar to purification schemes for CHO-derived biologics, many analytical parameters also have to be registered for the purification of plant-derived products. Despite the fact that molecular farming involves plants that are part of our normal diet, the injection of unwanted co-purifying plant constituents into humans is likely to provoke totally different and even fatal responses. Therefore, good manufacturing practice (GMP) would require the quantification of endotoxins and pesticide residuals, the removal of any co-purifying plant proteins and metabolites, determination of the extent of product aggregation, stability in final formulations, inactivation of eventual viral contaminations, and particle load. The costs for these measures can easily double production costs and should be included in the cost calculations of molecular farming when compared with classical fermentation.

The pharmaceutical industry anticipates that molecular farming will save time and money compared to traditional production systems. Because of bottlenecks and production costs, many biologics will never reach the market and the intended patients, or will do so only with great delays, if molecular farming fails. However, a number of points in the production of plant-derived proteins have yet to be addressed appropriately. In order to fulfill all requirements and obtain regulatory approval, the questions outlined above have to be answered for each recombinant protein. Last but not least, economical factors will decide whether molecular farming in plants will increase the number of available products.

12

BIOEMULSION

An emulsion is a heterogeneous preparation composed of two immiscible liquids (by convention described as oil and water), one of which is dispersed as fine droplets uniformly throughout the other. Emulsions are thermodynamically unstable and revert back to separate oil and water phases by fusion or coalescence of droplets unless kinetically stabilized by a third component, the emulsifying agent. The phase present as small droplets is called the disperse, dispersed, or internal phase and the supporting liquid is known as the continuous or external phase. Droplet diameters vary enormously, but in pharmaceutical emulsions they are typically polydispersed with diameters ranging from approximately 0.1 to 50 μm. Emulsions are conveniently classified as oil-in-water (o/w) or water-in-oil (w/o), depending on whether the continuous phase is aqueous or oily. Practical pharmaceutical emulsions, however, are rarely simple two-phase oil and water preparations; many are multicomponent systems containing additional solid or liquid crystalline (e.g., lamellar) phases. Multiple emulsions, which are prepared from oil and water by the reemulsification of an existing emulsion so as to provide two dispersed phases, are also of pharmaceutical interest. Multiple emulsions of the oil-in-water-in-oil (o/w/o) type are w/o emulsions in which the water globules themselves contain dispersed oil globules; conversely, water-in-oil-in-water (w/o/w) emulsions are those where the internal and external aqueous phases are separated by the oil. These more complex emulsions are covered by the broader International Union of Pure and Applied Chemistry (IUPAC) definition of emulsions, which extends the classical definition to include "liquid droplets and/or liquid crystals dispersed in a liquid."

Emulsions are formulated for virtually all the major routes of administration, and there are a number of dermatological, oral and, parenteral preparations available commercially. The internal phase may contain water-soluble drugs, preservatives, and flavoring agents whilst the oil phase may itself be therapeutically active or may act as a carrier for an oil-soluble drug. Such preparations provide an effective approach to many of the problems in drug delivery, often showing distinct advantages over other dosage forms by way of improved bioavailability and/or reduced side effects. However, despite such advantages, emulsions are not used as extensively as other oral or parenteral dosage forms due to the fundamental problems of emulsion instability that result in unpredictable drug release profiles and possible toxicity. The full potential of emulsions will not be realized until stable systems are developed with predictable in vitro and in vivo release patterns. Much of the emulsion research over the past decade is based on attempts to understand the relationships between emulsion stability, physicochemical properties, and biological fate. Multiple emulsions are even more difficult to stabilize, and characterize and although there is an increasing interest in their potential applications for drug delivery, at present there are no commercial preparations available.

Pharmaceutical Applications

The current and potential pharmaceutical applications of emulsions have been the subject of a number of general reviews. Traditionally the term "*emulsion*" is restricted to mobile emulsions for internal use; emulsions for external use are described by their pharmaceutical types as liniments, lotions, and creams. This tends to conceal the fact that by far the largest group of emulsions currently used in pharmacy and medicine are dermatological emulsions for external use. Both oil-in-water and water-in-oil emulsions are extensively used for their therapeutic properties and/or as vehicles to deliver drugs and cosmetic agents to the skin. The emulsion facilitates drug permeation into and through the skin by its occlusive effects and/or by the incorporation of penetration-enhancing components. Particular attention is paid to patient acceptance of such formulations, which range in consistency from mobile liniments and lotions to semisolid ointments and creams. In the past, the development of dermatological emulsions was essentially empirical with only a limited understanding of the underlying principles. Today, although the microstructure of many of these complex formulations is now better understood, the mechanisms by which the structure of an emulsion can influence drug bioavailability are far from clear and much of the literature on the role of emulsions in drug release to the skin is contradictory. Confusion arises because the majority of investigations concerning in vitro vehicle effects on drug release are only on bulk formulation. As most emulsions are applied to the skin as a thin film, the drug delivery system is not one of bulk emulsion, but rather a dynamic evaporating system in which phase changes can occur as the preparation is rubbed into the skin and the relative concentrations of volatile ingredients alter. Droplet size appears to influence drug delivery to the skin, with submicron lipid emulsions enhancing the transcutaneous permeation and efficiency of a number of lipophilic drugs.

Oral emulsions are almost exclusively of the oil-inwater type. They provide a degree of taste masking as the aqueous external phase effectively isolates the oil from the tongue. Mineral and castor oils have been emulsified in water and administered orally for the local treatment of constipation for many years as have various nutritional oils from fish liver (generally halibut or cod) or vegetable origin to produce oral liquid food supplements. It has long been established that the use of o/w emulsions as carriers for lipophilic drugs may improve oral bioavailability and efficacy. For example, griseofulvin formulated as an o/w emulsion has enhanced gastrointestinal absorption when compared with suspensions, tablets, or capsule dosage forms. The mechanisms by which emulsions modify and improve absorption processes are complex and not fully understood, although the oil itself influences gastric motility. Fats and oils are solubilized by the bile salts so that the administration of already emulsified oil droplets containing a high concentration of drug may increase the likelihood of further droplet and drug solubilization and transport across the GI tract by the fat absorption pathways.

The type of emulsion used parenterally depends on the route of injection and the intended use. Oil-in-water emulsions are administered by all the major parenteral routes whereas water-in-oil emulsions are generally reserved for intramuscular or subcutaneous administration where sustained release is required. Drug action is prolonged in such oily emulsions because the drug has to diffuse from the aqueous dispersed phase through the oil-continuous environment to reach the tissue fluids. Water-in-oil emulsions are used to disperse water-soluble immunizing antigens in mineral oil for injection via subcutaneous or intramuscular routes as adjuvant preparations where they prolong and enhance the antigenic stimulus and increase the antibody titer. Oily emulsion formulations also show promise in cancer chemotherapy as vehicles for prolonging drug release after intramuscular or intratumoral injection, and as a means of enhancing the transport of anticancer agents via the lymphatic system. Water-in-oil emulsions for sustained release are often difficult to inject because of the high viscosities of the oily continuous phases. Although these problems can be overcome by reemulsification of the primary w/o emulsion to produce a less viscous multiple w/o/w emulsion, a study using 5-fluorouracil implied that sustained release was actually less marked with multiple emulsions.

Sterile parenteral oil-in-water emulsions have been used extensively for over 40 years for the intravenous administration of fats, carbohydrates, and vitamins to debilitated patients. Several vegetable oil-in-water emulsions are now available commercially with droplet sizes similar to that of chylomicrons (approximately 0.5–2 μm), the natural fat droplets in the blood that transport ingested fats to the lymphatic and circulatory systems. More recently, such emulsions have been employed as intravenous carriers for poorly water-soluble lipophilic drugs such as vitamin K diazepam (e.g., Diemuls), vitamin A (Vitlipid N), and profonol (Diprovan) as alternatives to the traditionally used cosolvent, surfactant solubilized, or pH controlled parenteral solutions. The drug dissolved in the oil phase of the emulsion is unlikely to precipitate and cause pain when diluted by blood on injection, and if susceptible to hydrolysis or oxidation, it will be protected by the non-aqueous environment. Emulsion formulations of diazepam and more recently clarithromycin have been clinically shown to be less painful than solubilized preparations while emulsions containing amphoteracin B are less toxic. This emulsion was also shown to be an equally effective, cheaper, and more elegant alternative to a liposomal system. The enormous literature on the potential of lipid emulsions for drug delivery and targeting is discussed in a recent book.

Radiopaque emulsions, which have long been used as contrast media in conventional X-ray examinations of body organs, are finding further application with more sophisticated techniques including computed tomography, ultrasound, and nuclear magnetic resonance. Perfluorochemical emulsions are used as artificial blood substitutes. The potential advantages of such systems over donated blood are enormous with the elimination of major donor associated problems such as blood group incompatibilities and blood disease. The first commercial product, Fluosol-DA was licensed several years ago in a number of countries to reduce myocardial ischaemia in patients undergoing angioplasty; however, Fluosol-DA was not a commercial success due to its slow excretion rate and to its marked instability, which meant that it had to be stored in the frozen state. In addition,

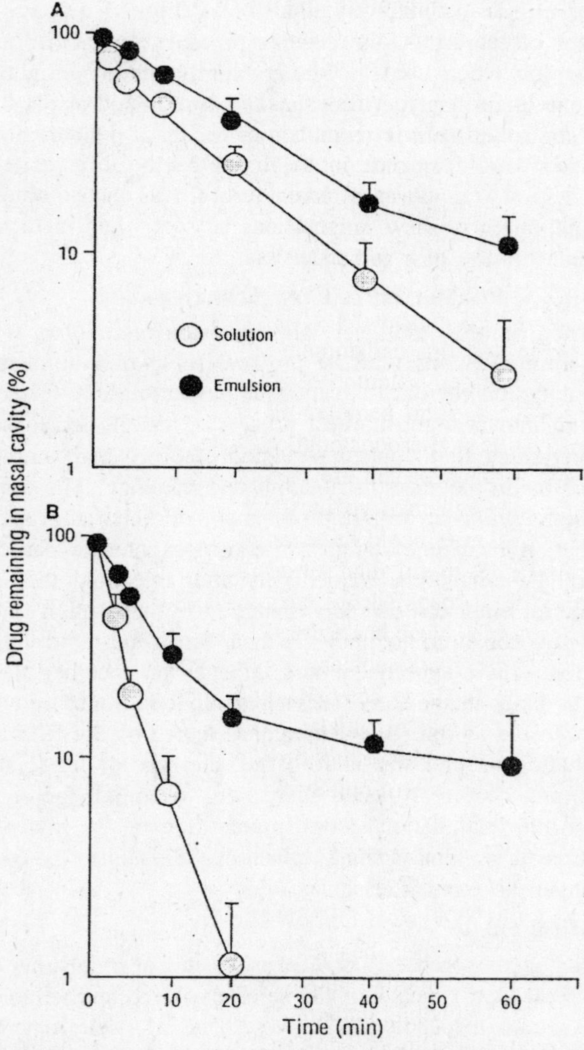

Fig. 12.1. Disappearance profiles of (A) tetrahydrozoline hydrochloride and (B) chlorpheniramine maleate from rat nasal cavity after nasal administration of an o/w emulsion and an aqueous solution at pH 8.

some patients were sensitive to one of the emulsifiers, pluronic F68. Currently, a second generation of emulsions is being evaluated to resolve the problems encountered with Fluosol.

There are only a few studies on the ocular and nasal applications of emulsions. Lipid (submicron) emulsions exhibited a long-lasting antidepressant effect on the intraocular pressure of rabbits after a single application when used as carriers for lipophilic antiglaucoma drugs. Medium-chain triglyceride emulsions formulated at pH 8 show potential as controlled release formulations for nasal delivery for they give prolonged drug residence in the nasal cavity. Enhanced nasal delivery of insulin was observed when insulin was incorporated into the continuous phase of an o/w emulsion, but not when incorporated into the aqueous phase of a w/o emulsion.

Formulation Considerations

The choice of oil, emulsifier, and emulsion type (o/w, w/o, or multiple) is limited by its ultimate use and route of administration. Potential toxicity and chemical incompatibilities in the final formulation must be taken into account as must processing details for these also affect the variables that control emulsion stability and therapeutic response such as droplet size distributions and rheology. The design of stable emulsions with the correct pharmacokinetic characteristics and tissue distribution is currently an area of enormous interest, particularly for parenteral IV emulsions. Immediately after injection, the surface of the parenteral emulsion droplets is altered by adsorption of blood components (optosonation) and they are then distributed rapidly through the circulation. Their subsequent fate depends on whether they are treated by the body in the same manner as chylomicrons, or whether they are recognized as foreign particles and cleared by the RES. Many factors, including droplet size and charge, the type of lipid, and the emulsifier composition influence their fate. A major factor to be considered in the formulation of oral preparations is the low pH and high ionic strength of stomach fluids, which may destabilize the emulsion by its effect on the emulsifier.

Pharmaceutical Oils

Oils used in the preparation of pharmaceutical emulsions are of various chemical types, including simple esters, fixed and volatile oils, hydrocarbons, and turpenoid derivatives. The oil itself may be the medicament, it may function as a carrier for a drug, or even form part of a mixed emulsifier system as in the case of some fixed oils that contain sufficient free fatty acids. Many oils, particularly those of vegetable origin, are liable to autooxidation with subsequent rancidity, and it is frequently necessary to add an antioxidant and/or preservative to inhibit this degradation process. For externally applied emulsions, mineral oils, either alone or combined with soft or hard paraffins, are widely used both as the vehicle for the drug and for their occlusive and sensory characteristics. The most widely used oils in oral preparations are non-biodegradable mineral and castor oils that provide a local laxative effect, and fish liver oils or various fixed oils of vegetable origin as nutritional supplements.

The choice of oil is severely limited in emulsions for parenteral administration for reasons of toxicity. Purified soybean, sesame, safflower, and cottonseed oils composed mainly of long-chain triglycerides have been used for many years as they are resistant to rancidity and show few clinical side effects. More recently, it has been recognized that the structure of the oil will influence the fate of emulsion droplets after injection. Mixtures containing both long- and medium-chain triglycerides are not only better energy sources for nutritional purposes but they are also cleared more rapidly from the circulation; such mixtures are now used in commercial preparations. Structured triglycerides, formed by modifying the oil enzymatically to produce 1,3-specific triglycerides are an area of increasing interest because of their influence on the in vivo circulation time of an emulsion. Purified mineral oil is used in some water-in-oil depot preparations where mineral toxicity must be carefully balanced against efficiency. Emulsified perfluoro-chemicals are considered acceptable for IV use provided that they are excreted relatively fast. A major problem in the formulation of the early perfluorocarbon emulsions was that the oils that form the most stable emulsions were not cleared rapidly from the body.

Pharmaceutical Emulsifiers

Emulsifying agents are used both to promote emulsification at the time of manufacture and to control stability during a shelf life that can very from days for extemporaneously prepared emulsions to months or years for commercial preparations. In practice, combinations of emulsifiers rather than single agents are used. The emulsifier also influences the in vivo fate of lipid parenteral emulsions by its influence on the surface properties of the droplets and on the droplet size distributions. For convenience, most pharmacy texts classify emulsifiers into three groups: (i) surface active agents; (ii) natural (macromolecular) polymers; and (iii) finely divided solids.

Surface Active Agents

Surfactants are manufactured from a variety of natural and synthetic sources and consequently they show considerable batch-to-batch variations in their homologue compositions and in trace impurities from the starting material. For example, batch variations in the number of neutral phospholipids occur in lecithin surfactants and non-ionic polyethylene surfactants show variations in the number of moles of ethylene oxide. The mechanisms by which such batch variations lead to differences in emulsifying properties are now better understood.

Although synthetic and semisynthetic surfactants form by far the largest group of emulsifiers studied in the scientific literature and many of them are available commercially, their use in pharmaceutical emulsions is limited by the fact that the majority are toxic (i.e., haemolytic) and irritant to the skin and mucous membranes of the gastrointestinal tract. In general, cationic surfactants are the most toxic and irritant and non-ionic surfactants the least. Surfactants are therefore used mainly at relatively low concentrations in topical preparations. The quaternary ammonium compounds constitute an important group of cationic emulsifiers in dermatological preparations because they have antimicrobial properties in addition to their o/w emulsifying action. There are many non-ionic surfactants with different oil and water solubilities available commercially because for each fatty starting material the polyoxyethylene chain length can be modified by the systematic addition of ethylene oxide groups. However, a limited number of polysorbate surfactants are used in oral emulsions, and parenteral preparations appear to be based only on the lecithins from plant or animal sources and the non-ionic polyoxyethylene oxide/ polyoxypropylene oxide block copolymer poloxomer 188, although some patients using the first generation of perfluorochemical emulsions were sensitive to this poloxomer. The emulsifier influences both emulsion stability and in vivo disposition by its influence on droplet surface properties.

Natural macromolecular materials and finely divided solids

Materials derived from natural sources may originate from animal or vegetable sources and many of these products are susceptible to degradation. For example, depolymerization (the polysaccharides) or hydrolysis (the steroids) usually lead to loss in emulsifying power. Some of these materials, polysaccharides and proteins in particular, provide good culture medium for microorganisms, and therefore preservation of emulsions containing them is imperative. To overcome these problems, a number of purified and semisynthetic derivatives are available, including various purified wool fat derivatives and semisynthetic celluloses such as methylcellulose and sodium carboxymethylcellulose. These are generally more stable than the unmodified materials. These celluloses are used in oral preparations; they are less suitable for topicals because of their unpleasant feel. Finely divided solids such as clays are used in dermatological preparations as structuring agents.

Preservatives

It is essential that emulsions are formulated to resist microbial attack, as this not only can affect the physicochemical properties of the formulation, causing color, odor, or pH changes and even phase separation, but may also constitute a health hazard. The potential sources of contamination can be from raw materials (especially if these are natural products), water, manufacturing and packaging equipment, or patients themselves. W/o emulsions are less susceptible to attack than o/w emulsions because the aqueous continuous external phase can produce ideal conditions for the growth of bacteria, moulds, and fungi. Preservatives are not used in parenteral emulsions, which are sterilized, generally by autoclaving, but sometimes by using sterile components and aseptically assembling the final emulsion.

There is no simple way of predicting the ideal preservative for a particular emulsion. In addition to requiring a wide spectrum of activity against bacteria, yeasts, and molds, the preservative should be free from toxic, irritant, or sensitizing activity. Some commonly used preservatives in oral and topical preparations include phenoxyethanol, benzoic acid, parabenzoates, and chlorcresol. Emulsions are heterogeneous products, and the preservative partitions between the oil and aqueous phases. As a sufficient aqueous concentration of the active (usually unionized) form must be present to ensure proper preservation, pH is an additional factor to be considered. Problems often arise because many of the materials used in emulsion formulation, for example hydrocolloids or polyoxyethylene surfactants, can interact with the preservatives, thus depleting their activity. The use of a single preservative is often considered unrealistic, and attention is being increasingly focused on the use of mixtures for a wider spectrum of activity, although this may introduce additional compatibility problems.

Antioxidants and Humectants

Antioxidants are added to many pharmaceutical preparations to prevent oxidative deterioration on storage of the oil, emulsifier, or the drug itself. Such deterioration, as well as destabilizing the formulation, imparts an unpleasant odor or taste. Some oils are supplied containing antioxidants already. Those commonly used in pharmacy include *butylated hydroxyanisole* (BHA) and *butylated hydroxytoluene* (BHT) at concentrations up to 0.2%, and the alkyl gallates. Humectants such as propylene glycol, glycerol, and sorbitol (5%) are often added to dermatological preparations to reduce the evaporation of water from the emulsion during storage and use. They are sometimes claimed to prevent evaporation of water from the surface of the skin, although their use at high concentrations would be expected to have the opposite effect (i.e., remove moisture and dehydrate the skin).

Emulsion Formation

There are essentially two major considerations in emulsification: first, the formation of emulsions of the correct type, oil-in-water, water-in-oil, or multiple emulsion with the required droplet size distributions and second, the stabilization of the dispersed droplets so formed. When given amounts of two immiscible liquids are mixed or mechanically agitated in the absence of other additives, both phases tend to form droplets of various sizes. The size distributions are related to the forces involved during the agitation process, and the number of droplets of each liquid depends on its relative volume. The surface free energy of the system, which is dependent on both total surface area and interfacial tension is raised by the increase in surface area produced during dispersion, and the system is thermodynamically unstable. To reduce this, high- energy droplets first assume a spherical shape, as this gives the minimum surface area for a given volume, and then on collision the droplets will coalesce (fuse) to reduce the interfacial area, the interfacial tension remaining constant.

The type of emulsion that forms, either o/w or w/o, depends on the relative rates of coalescence of each type of droplet, with the more rapidly coalescing droplets forming the continuous phase. Generally, this is the liquid present in the larger amount because higher number of droplets increase the probability of collision and coalescence. With phase volumes of oil and water close to 50%, other factors such as the order and rate of addition of each liquid are important. If agitation ceases, coalescence will continue until complete phase separation—the state of minimum free energy—is reached. Thus, emulsification can be considered as the result of two competing processes, namely the disruption of bulk liquids to produce fine droplets and the recombination of the droplets to give back the original bulk liquids. With the inclusion of surfactants or other classes of emulsifiers, the type of emulsion that forms is no longer simply a function of the phase volume and the order of mixing, but also the relative solubilities of the emulsifier in oil and water. In general, the phase in which the emulsifier is most soluble becomes the continuous phase (Bancroft's Rule); thus, hydrophilic polymers and surfactants promote o/w emulsions whereas lipophilic surfactants promote w/o emulsions. Preferential coalescence of the phase in which the emulsifier is most soluble occurs because when droplets collide, the emulsifier is easily displaced from the interface into the droplet, thus providing little protection against coalescence. Theoretically, the disperse phase of an emulsion can occupy up to 74% of the phase volume, and such high internal phase o/w emulsions have been produced with suitable surfactants. It is more difficult to formulate w/o emulsions with greater than 50% disperse phase because of the steric mechanisms involved in their stabilization (discussed later), and the addition of extra water sometimes causes inversion to an o/w emulsion.

Emulsion Characteristics

In general, an emulsion exhibits the characteristics of its external phase. Several methods are available for identifying the emulsion type. Dilution tests are based on the fact that the emulsion is only miscible with the liquid that forms its continuous phase. Conductivity measurements rely on the poor conductivity of oil compared with water, and give low values in w/o emulsions where oil is the continuous phase. Staining tests in which a water-soluble dye is sprinkled onto the surface of the emulsion also indicate the nature of the continuous phase. With an o/w emulsion there is rapid incorporation of the dye into the system whereas with the w/o emulsion the dye forms microscopically visible clumps. The reverse happens on addition of an oil-soluble dye. These tests essentially identify the continuous phase and do not indicate whether a multiple emulsion has been produced. This can be resolved by microscopy.

Rheology

The rheological properties of emulsions are influenced by a number of interacting factors, including the nature of the continuous phase, the phase volume ratio, and to a lesser extent, particle size distributions. A variety of products ranging from mobile liquids to thick semisolids can be formulated by altering the dispersed phase volume and/or the nature and concentration of the emulsifiers. For low internal phase volume emulsions, the consistency of the emulsion is generally similar to that of the continuous phase; thus, w/o emulsions are generally thicker than o/w emulsions, and the consistency of an o/w system is increased by the addition of gums, clays, and other thickening agents that import plastic or pseudoplastic flow properties. Some mixed emulsifiers interact in water to form a viscoelastic continuous phase to give a semisolid o/w cream.

Droplet Size Distributions

Droplet size distributions in pharmaceutical emulsions are important from both stability and biopharmaceutical considerations. The larger the particle size, the greater the tendency to coalesce and further increase droplet size. Thus, fine particles generally promote better stability. Size distributions are influenced by the characteristics of the emulsifier and the method of manufacture. From a biological point of view, fine emulsification enhances gastrointestinal absorption and whilst this is desirable with oral formulations of nutrient oils alone or with drugs dissolved in them, it may give adverse clinical effects with mineral oils that are used for a local effect and are toxic if absorbed. Droplets in emulsions used as contrast media incomputed tomography are approximately 1–3 μm. Parenteral emulsions should be formulated so that the dispersed droplet sizes range from approximately 100 nm to 1 μm. In any event, sizes should never be greater than 5 μm in diameter because of the danger of pulmonary emboli. There is clear evidence that, as with other colloidal preparations, droplet size distributions influence the clearance kinetics of parenteral emulsions. Larger droplets are treated as foreign bodies and rapidly cleared by elements of the RES while smaller droplets may be treated as natural fat sorting lipoproteins, with a different in vivo fate. Drug delivery from dermatological preparations also appears to be improved in lipid emulsions containing submicron droplets.

Emulsion Stability

A stable emulsion is considered to be one in which the dispersed droplets retain their initial character and remain uniformly distributed throughout the continuous phase for the desired shelf life. There should be no phase changes or microbial contamination on storage, and the emulsion should maintain elegance with respect to odor, color, and consistency. Instabilities of both chemical and physical origins can occur in emulsion formulations. Chemical instabilities, such as the development of rancidity in natural oils due to oxidation by atmospheric oxygen, the depolymerization of macromolecular emulsifiers by hydrolysis, or microbial degradation can be minimized by the addition of suitable antioxidants and preservatives. More general chemical instabilities involving interactions between the drug and emulsion excipients or between the excipients themselves may lead to physical instabilities. If these interactions involve the emulsifying agent, they may destroy its emulsifying properties, causing the emulsion to break. For example, interactions between phenolic preservatives and polyoxyethylene non-ionic emulsifiers may lead to loss of emulsifying power as well as poor preservation.

Physical Instability

As emulsions are inherently unstable, they eventually revert to the original state of two separate liquids, that is, will break or crack. In the presence of an emulsifier and other additives, this state is approached via several distinct processes, some of which are reversible

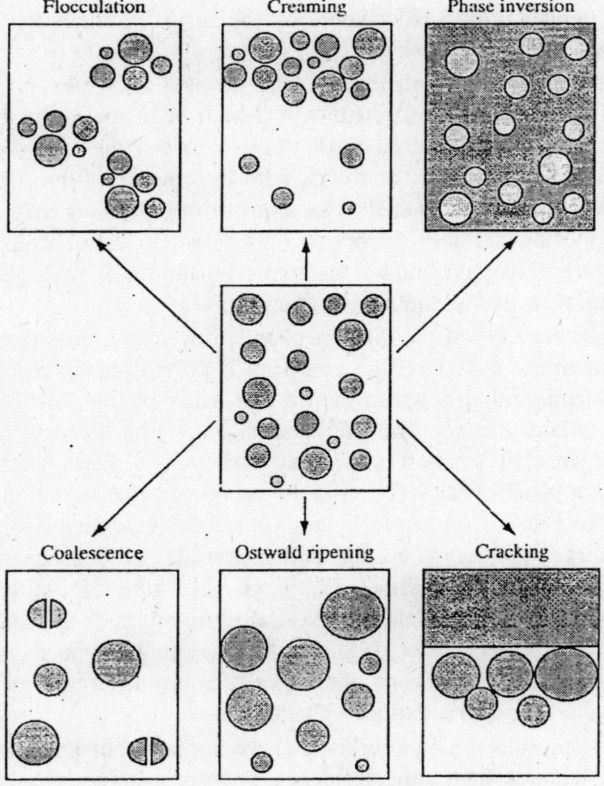

Fig. 12.2. Schematic representation of the various processes of emulsion breakdown.

such as creaming and flocculation and others irreversible such as coalescence and Ostwald ripening. Phase inversion when an oil-in-water emulsion inverts to form a water-in-oil emulsion or visa versa is a special case of irreversible instability that occurs only under well-defined conditions such as a change in emulsifier solubility due to specific interactions with additives or to a change in temperature.

Flocculation describes a weak reversible association between emulsion globules separated by thin films of continuous phase. The individual droplets retain their separate identities, but each floccule or cluster of droplets behaves physically as a single unit. The association arises from the interaction of attractive and repulsive forces between droplets and is reversible in the sense that the original dispersion can generally be regained by mild agitation. Flocculation is generally regarded as a precursor to the irreversible process of coalescence, although sometimes the time scale between flocculation and coalescence can be extended almost indefinitely by the adsorbed emulsifier, giving a kinetically stable emulsion.

Coalescence, where dispersed phase droplets merge to form larger droplets, takes place in two distinct stages. It begins with the drainage of liquid films of continuous phase from between the oil droplets as they approach one another and ends with the rupture of the film when a critical thickness is reached. The approaching droplets may deform as the opposing surfaces distort to either flatten (small droplets) or dimple (larger droplets) under the hydrodynamic pressures generated by viscous flow of the continuous phase.

Coalescence is not the only mechanism by which dispersed phase droplets increase in size. If the emulsion is polydispersed and there is significant miscibility between the oil and water phases, then Ostwald ripening, where droplet sizes increase due to large droplets growing at the expense of smaller ones, may also occur. This destabilizing process is a result of the Kelvin effect and occurs when small emulsion droplets (less than 1 μm) have higher solubilities (and vapor pressures) than do larger droplets (i.e., the bulk material) and consequently are thermodynamically unstable. To reach the state of equilibrium, molecules from these droplets dissolve and diffuse through the continuous phase to enlarge the larger droplets. As the small droplets lose their oil, they become even smaller, the vapor pressure difference increases, and Ostwald ripening is further enhanced.

Creaming or sedimentation occurs when the dispersed droplets or floccules separate under the influence of gravity to form a layer of more concentrated emulsion, the cream. Generally a creamed emulsion can be restored to its original state by gentle agitation. This process, which inevitably occurs in any dilute emulsion if there is a density difference between the phases as a consequence of Stokes law, should not be confused with flocculation which is due to particle interactions resulting from the balance of attractive and repulsive forces. Most oils are less dense than water so that the oil droplets in o/w emulsions rise to the surface to form an upper layer of cream. In w/o emulsions, the cream results from sedimentation of water droplets and forms the lower layer. According to Stokes Law, the rate of creaming can be minimized by reducing droplet sizes and/or thickening the continuous phase. Adjustment of the densities of the two phases has received little attention.

The destabilization processes are not independent and each may influence or be influenced by the others. For example, the increased droplet sizes after coalescence or Ostwald ripening will enhance the rate of creaming, as will the formation of large floccules which behave as single entities. In practice, creaming, flocculation, and Ostwald ripening may proceed simultaneously or in any order followed by coalescence.

Coalescence and Ostwald ripening are obviously the most serious types of instability as they result in the formation of progressively larger droplets and ultimately lead to phase separation. Creaming and flocculation, on the other hand, are more subtle forms of instability, for although they represent potential steps towards coalescence and breaking due to the close proximity of the droplets, many practical emulsions remain in this state for long periods of time without significant coalescence and can be redispersed simply by shaking the container.

Emulsion Stabilization

Emulsifiers stabilize emulsions in a number of different ways, all of which act to prevent or delay the various destabilization processes described previously. The emulsifier may form an interfacial film at the oil–water interface and/or structure (i.e., thicken) the continuous phase. The interfacial film introduces additional repulsive (e.g., electrostatic, steric, or hydrational) forces between droplets to counteract attractive van der Waals forces and inhibit the close approach of droplets. It may also provide a barrier to the coalescence of droplets in close proximity, particularly if the film is close-packed and elastic. Surfactant interfacial films also lower the interfacial tension between oil and water. Although this effect is important during the emulsification process where it facilitates the breakup of droplets, it is not a major factor inmaintaining the long-term stability of emulsions. Inemulsions that are thickened by the emulsifier, the interfacial film does not play the dominant role in maintaining stability; rather, it is the structured continuous phase that forms a rheological barrier to prevent the movement and hence the close approach of droplets and also inhibits Ostwald ripening.

Classical (Interfacial) Theories

Classical theories of emulsion stability focus on the manner in which the adsorbed emulsifier film influences the processes of flocculation and coalescence by modifying the forces between dispersed emulsion droplets. They do not consider the possibility of Ostwald ripening or creaming nor the influence that the emulsifier may have on continuous phase rheology. As two droplets approach one another, they experience strong van der Waals forces of attraction, which tend to pull them even closer together. The adsorbed emulsifier stabilizes the system by the introduction of additional repulsive forces (e.g., electrostatic or steric) that counteract the attractive van der Waals forces and prevent the close approach of droplets. Electrostatic effects are particularly important with ionic emulsifiers whereas steric effects dominate with non-ionic polymers and surfactants, and in w/o emulsions. The applications of colloid theory to emulsions stabilized by ionic and non-ionic surfactants have been reviewed as have more general aspects of the polymeric stabilization of dispersions.

The DLVO theory, which was developed independently by Derjaguin and Landau and by Verwey and Overbeek to analyze quantitatively the influence of electrostatic forces on the stability of lyophobic colloidal particles, has been adapted to describe the influence of similar forces on the flocculation and stability of simple model emulsions stabilized by ionic emulsifiers. The charge on the surface of emulsion droplets arises from ionization of the hydrophilic part of the adsorbed surfactant and gives rise to electrical double layers. Theoretical equations, which were originally developed to deal with monodispersed inorganic solids of diameters less than 1 μm, have to be extensively modified when applied to even the simplest of emulsions, because the adsorbed emulsifier is of finite thickness and droplets, unlike solids, can deform and coalesce. Washington has pointed out that in lipid emulsions, an additional repulsive force not considered by the theory due to the solvent at close distances is also important.

The theory states that the forces between droplets can be considered as the sum of an attractive van der Waals part V_A and a repulsive electrostatic part V_R when identical electrical double layers overlap. As the origin of each force is independent of the other, each is evaluated separately, and the total potential of interaction V_T between the two droplets as a function of their surface-to-surface separation is obtained by summation

$$V_T = V_A + VR$$

It can be seen that a weak attraction occurs at large droplet separations represented by the secondary energy minimum, and a very strong attraction at small droplet separations hence the very deep primary minimum. At intermediate distances, double-layer repulsion dominates and there is a maximum in the curve. Flocculation occurs in the secondary minimum, where the attractive forces are relatively weak and floccules are easily separated by low energy agitation. Once flocculated, droplets are prevented from approaching closer by the potential energy barrier. If they have sufficient energy to overcome the barrier, the process of coalescence commences as the droplets move closer together. Once in the primary minimum the aggregates formed are separated by only a small distance so that stability against coalescence is determined by the resistance of the interfacial film to rupture.

The height of the energy barrier, which is crucial to emulsion stabilization, depends on the state of ionization of the emulsifying agent. Most surfactants are used at pH values where they are totally ionized so that the surface potential is high, giving a correspondingly high energy barrier. The surface potential cannot be measured directly, but can be estimated from the experimentally derived zeta potential. In lipid emulsions for parenteral nutrition, the electrostatic barrier is provided by the ionization of the negatively charged phospholipids in the emulsifier film at the oil droplet–water interface. At physiological pH, a typical fat emulsion carries a negative charge with the zeta potential between 30 and 60 mV. This is sufficient to ensure stability because of the high potential energy barrier. The addition of electrolytes or a change in pH can have a devastating effect on emulsion stability by compressing the double layers, thus reducing the zeta potential and energy barrier and allowing droplets to move into the primary minimum. Thus, great care must be exercised when electrolytes are added nutritional emulsions. With emulsifiers such as proteins and gums, ionization, and hence emulsifying activity, is also pH dependent.

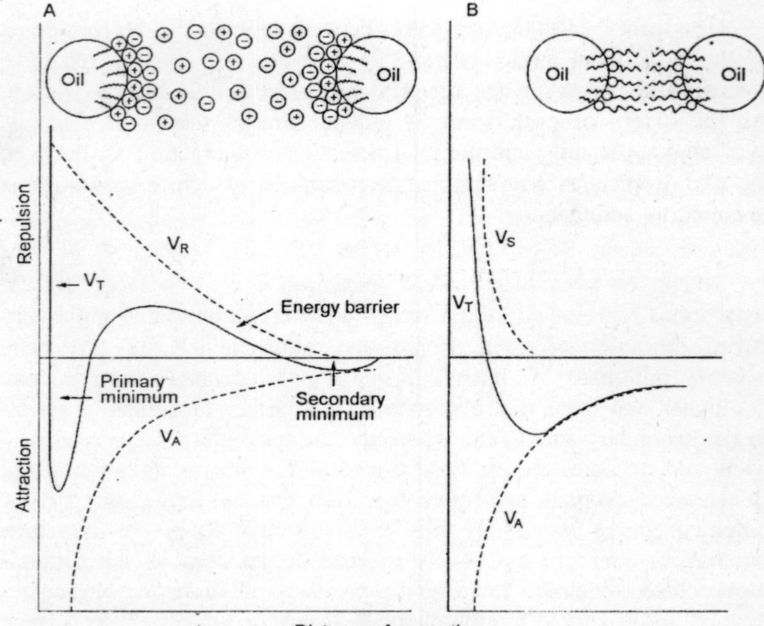

Fig. 12.3. The total potential energy of interaction V_T as a function of distance of surface separation H for two similar oil droplets in an oil-in-water emuslion.

The DLVO theory does not explain either the stability of water-in-oil emulsions or the stability of oil-in-water emulsions stabilized by adsorbed non-ionic surfactants and polymers where the electrical contributions are often of secondary importance. In these, steric and hydrational forces, which arise from the loss of entropy when adsorbed polymer layers or hydrated chains of non-ionic polyether surfactant intermingle on close approach of two similar droplets, are more important. In emulsions stabilized by polyether surfactants, these interactions assume importance at very close distances of approach and are influenced markedly by temperature and degree of hydration of the polyoxyethylene chains. With block copolymers of the ethylene oxide–propylene oxide type, such as the poloxamers, the hydrated polyoxyethylene chains extend into the continuous phase to provide steric stabilization and the hydrophobic propylene oxide portion is anchored onto the droplet surface to form a strong protecting layer against coalescence. Stability is optimized when the droplet surfaces are completely coated by polymer chains so that desorption and lateral movement of the polymer is inhibited. With w/o emulsions, steric hindrance of the adsorbed chains of emulsifier can also result in entropic repulsion effects at small distances of separation.

Some natural polymeric emulsifiers such as the gums, in addition to forming steric and electrostatic barriers form thick multilayered films that are very resistant to film rupture. They may also thicken the continuous phases of o/w emulsions, thereby reducing the rate of film drainage in the initial stages of coalescence. Small solid particles may stabilize emulsions if they are wetted by both phases and possess sufficient adhesion for one another to form a coherent interfacial film. The film serves as a mechanical barrier to prevent the coalescence of droplets, and if charged, electrostatic mechanisms further assist in the stabilization of the emulsion. Although solids are not generally sufficient to stabilize emulsions on their own, they often reinforce the effectiveness of other emulsifiers.

Stabilization by Mixtures of Emulsifiers

Most pharmaceutical emulsions, whether dilute mobile systems for internal use or thick semisolid creams for application to the skin, contain mixtures of emulsifiers, as these provide more stable preparations. For example, traditional oral preparations are sometimes stabilized by mixtures of gums such as acacia and tragacanth and mixtures of non-ionic surfactants of high and low *hydrophile–lipophile balance* (HLB) generally form more stable emulsions than a single surfactant. The lecithins used to stabilize parenteral emulsions are usually mixtures of neutral and charged lipids as are the partially neutralized glyceryl esters such as self-emulsifying glyceryl monostearate. Combinations of sparingly soluble long-chain acids, alcohols, or glyceryl esters with more soluble ionic and non-ionic surfactants are widely used in dermatological o/w lotions and creams, where they are sometimes added in the form of a preblended emulsifying wax. Surfactant/fatty acid combinations are also present in traditional liniment and lotion emulsion formulations prepared by the nascent soap method and in preparations where triethanolamine soaps are formed in situ from the interaction of triethanolamine and excess fatty acid.

Equations from the DLVO theory even if modified to allow for the steric repulsive forces cannot cope with mixtures of emulsifiers. Increased stability in model emulsions is attributed not so much to the control of flocculation (although this does occur), but rather to the prevention or retardation of coalescence by closer packing of the molecules in the adsorbed monolayer to form a more rigid and condensed film. There is now substantial evidence that interactions between

Table 12.1. Typical emulsifying waxes and their component surfactants

Emulsifying wax	*Components*
Emulsifying wax USNF	Cetearyl alcohol, polysorbate
Cationic emulsifying wax BP	Cetearyl alcohol, cetrimonium bromide
Cetomacrogol emulsifying wax BP	Cetearyl alcohol, ceteth 20
Glyceryl stearate, SE	Glyceryl stearate, soap

emulsifier components to form specific lamellar phases, either liquid crystalline or gel, that are capable of incorporating large volumes of water are important for the stability of many parenteral and dermatological emulsions. Mobile parenteral injections stabilized by phospholipid mixtures usually contain swollen lamellar liquid crystals whereas a swollen gel phase which generally provides better stability as well as a means of controlling rheological properties dominates in semisolid dermatological emulsions prepared with emulsifying waxes. Much of the information about their structures was obtained from investigations of the phase behaviour of emulsifiers and their components in water over the ranges of concentration and temperature relevant to the manufacture, storage, and use of the formulations. It is interesting to note that the same electrostatic, hydrational, and steric forces that operate in simple emulsions also dominate the stability and properties of the lamellar phases.

Ostwald Ripening

Ostwald ripening has not been studied as extensively in emulsions as has coalescence, although it is a major mechanism for instability in lipid and perfluorochemical emulsions with submicron droplet sizes where a condensed monolayer is not always necessary for emulsion stability. Although surfactant interfacial films protect against flocculation and coalescence, Ostwald ripening may in fact be enhanced if the surfactant is above the *critical micelle concentration* (cmc) because of the diffusion of solubilized oil through the continuous phase. The addition of a third component to the emulsion that has a lower vapor pressure and solubility than the disperse phase will also inhibit Ostwald ripening. The addition of long-chain alkanes to comparatively unstable oil-in-water emulsions prepared with sodium dodecyl sulfate resulted in marked increases in stability even though the alkanes do not effect the composition or mechanical properties of the oil–water interface. The stability of pure perfluorodecalin emulsions used as blood substitutes is enhanced by the addition of a small quantity of perfluorotributylamine, and lipid emulsions containing local anaesthetic/analgesic drugs show enhanced stability in the presence of hydrophobic excipients of lower solubility than the disperse phase. Polymeric emulsifiers possibly stabilize emulsions against Ostwald ripening by increasing the viscosity of the continuous phase. The relative lack of Ostwald ripening in emulsions prepared from oils immiscible with water, such as mineral oil, may partly explain why they are easier to emulsify than are more miscible vegetable oils used in parenteral preparations.

Selection of Emulsifier

Over the years there have been many attempts to find systemic methods for screening potential emulsifiers from the enormous number of surfactants available commercially. Although the mechanisms governing the stability of emulsions, including the complex multiple phase systems of pharmacy are becoming clearer, there are still few scientific guidelines to assist in the proper selection of emulsifiers for a particular emulsion. Semiempirical methods based on both interfacial considerations and the phase behavior of the emulsifiers are considered briefly next.

Hydrophile–lipophile balance (HLB) concept

Griffin devised the concept of *hydrophile–lipophile balance* (HLB) and its additivity many years ago for selection of non-ionic emulsifiers and this rather empirical method is still widely used. Each surfactant is allocated an HLB number usually on a scale of 0–20, based on the relative proportions of the hydrophilic and hydrophobic part of a molecule. Water-in-oil emulsions are formed generally from oil-soluble surfactants of low HLB number and oil-in-water emulsions from more hydrophilic surfactants of high HLB number. The method of selection is based on the observation that each type of oil will require an emulsifying agent of a specific HLB number to produce a stable emulsion. Thus, oils are often designated two “required” HLB numbers, one low and one high, for their emulsification to form water-in-oil and oil-in-water emulsions respectively. A series of emulsifiers and their blends with HLB values close to the required HLB of the oil are then examined to see which one forms the most stable emulsion.

Although the HLB concept narrows the range of emulsifiers to select and provides a schematic approach for the formulator, it is limited by its strict relation to molecular structure of the individual surfactants. The concept does not consider the total emulsion and is therefore insensitive to interactions between emulsifier components, the influence of temperature changes, or the presence of additional ingredients in the emulsion. Consequently, not all emulsifier blends of the correct HLB form stable systems. For example, when surfactants of widely different HLB numbers are blended to give the optimum theoretical HLB, the high solubility of the surfactant in the oil and aqueous phases change the balance of the molecules at the interface and unstable emulsions may result. Similarly, if the added surfactants form intermolecular associations at the interface, the association complex is unlikely to have properties that are related in any simple way to the individual properties of the constituent molecules.

Phase inversion temperature (PIT) method (HLB-temperature)

A complementary means of emulsifier selection, the *phase inversion temperature* (PIT), which employs a characteristic property of the emulsion rather than the properties of the emulsifiers in isolation, was introduced by Shinoda. The method uses the fact that the stabilities of oil-in-water emulsions containing nonionic surfactants are closely related to the degree of hydration of the interfacial films. Emulsion stability is reduced by increase in temperature or added salts because these decrease the extent of interfacial film hydration. Phase inversion, due to a change from preferential water solubility of the emulsifier film at low temperature to preferential oil solubility at high temperature, will occur at a specific temperature unique to the particular emulsion and this can be determined experimentally. As a general rule, relatively stable oil-in-water emulsions are obtained when their temperatures during storage and use are between 20 and 65°C below the PIT, presumably because the films are sufficiently hydrated. Mixtures of emulsifiers with identical HLBs produce emulsions with quite different PITs because additives and interactions between the components affect PIT but not HLB.

Microscopic selection for multiple phase emulsions

The better understanding of the mechanisms of stability incomplex dermatological emulsions stabilized by surfactants and amphiphiles has enabled the development of a rapid microscopic method for evaluation of potential emulsifiers. The method is based on the observation that good emulsifier blends that stabilize emulsions by the formation of multilayers of stable gel phase also swell spontaneously in water at ambient temperature and this process can be observed microscopically. Mixtures that do not form gel phase or form metastable gels only after a heating and cooling cycle cannot be observed to swell spontaneously at ambient temperature.

Emulsification Techniques

Emulsions are usually prepared by the application of mechanical energy produced by a wide range of agitation techniques. These disrupt droplets by the application of either shear forces in laminar flow or inertial forces in turbulent flow. Emulsifying devices ranging from simple hand mixers and stirrers to the use of propeller or turbine mixers, static mixers, colloid mills, homogenizers, and ultrasonic devices have been used.

Emulsifiers also have an important role in the process of emulsification. Surfactant emulsifiers reduce interfacial tensions during emulsification, making droplets easier to break up as well as reducing the tendency for recombination. Other emulsifiers such as the polymer macromolecules alter the hydrodynamic forces during the agitation process by their influence on rheological properties. Scale-up procedures from the laboratory to manufacture can introduce a number of problems due to the difficulties in matching the exact conditions of mixing, and, because of entrapment-of air, especially in emulsions of high consistency that have a yield value. Along with being inelegant, even traces of atmospheric air can cause decomposition in drugs or excipients susceptible to oxidation.

There are additional constraints when manufacturing parenteral emulsions that must be sterile and of fine particle size. Perfluorochemical and fat emulsions are usually prepared by homogenization at high temperature and pressure, as a large output of energy is required to produce droplet sizes considerably less than 1 μm. Although heat sterilization is widely used, this places a severe test on the stability, and emulsions are sometimes prepared from sterile components under strict aseptic conditions and further sterilized by filtration.

Processing variables

Differences in manufacturing techniques such as the rate of the heating and cooling cycle, the extent and order of mixing can cause variations in the consistency and rheology of the resulting emulsions. The initial particle size of the emulsion depends on the emulsifiers used, the emulsification equipment, the addition speed, and the phase volume. If the surfactant is placed in one of the phases prior to emulsification, it will migrate to the other to establish equilibrium. Thus, emulsification temperatures and cooling rates are important and the time of the mixing should be sufficient to allow the surfactant to migrate to and equilibrate at the interface throughout the process. Oil-in-water emulsions are sometimes prepared by the phase inversion technique, where the aqueous phase is added to the oil phase to form a w/o emulsion that inverts to an o/w emulsion on addition of further amounts of water. This process is claimed to give finer emulsions.

Preparation techniques, in particular cooling rates and mixing procedures, have a marked effect on initial and final consistencies of emulsions prepared with nonionic emulsifying waxes. For example, "*shock*" cooling and limited mixing initially produces very mobile systems whereas slow cooling with adequate mixing produces semisolid emulsions. Mixing time, when the emulsifiers are in the molten state, influences the distribution of surfactant within the molten masses and bilayers and the relative lamellar order within the system. With ionic emulsifying waxes, different preparation techniques cause comparatively minor variations in the consistency of the final product. It was shown that differences are not due to the gel phase component of cationic ternary systems, but rather due to the variations in size of the crystalline alcohol that precipitates after manufacture. Systems formed by a rapid "*shock*" cooling method exhibited smaller but greater numbers of cetostearyl alcohol crystals and were thicker than similar ternary systems manufactured by a more lengthy procedure.

Microemulsion

Microemulsions are thermodynamically stable, transparent (or translucent) dispersions of oil and water that are stabilized by an interfacial film of surfactant molecules. The surfactant may be pure, a mixture, or combined with a cosurfactant such as a medium-chain alcohol (e.g., butanol, pentanol). These homogeneous systems, which can be prepared over a wide range of surfactant concentrations and oil to water ratios (20–80%), are all fluids of low viscosity.

The term microemulsion, which implies a close relationship to ordinary emulsions, is misleading because the microemulsion state embraces a number of different microstructures, most of which have little in common with ordinary emulsions. Although microemulsions may be composed of dispersed droplets of either oil or water, it is now accepted that they are essentially stable, single-phase swollen micellar solutions rather than unstable two-phase dispersions. Microemulsions are readily distinguished from normal emulsions by their transparency, their low viscosity, and more fundamentally their thermodynamic stability and ability to form spontaneously. The dividing line, however, between the size of a swollen micelle (~10– 140nm) and a fine emulsion droplet (~100–600nm) is not well defined, although microemulsions are very labile systems and a microemulsion droplet may disappear within a fraction of a second whilst another droplet forms spontaneously elsewhere in the system. In contrast, ordinary emulsion droplets, however small, exist as individual entities until coalescence or Ostwald ripening occurs.

At high water concentrations, microemulsions consist of small oil droplets dispersed in water (o/w microemulsion), while at lower water concentrations the situation is reversed and the system consists of water droplets dispersed in oil (w/o microemulsions). In each phase, the oil and water droplets are separated by a surfactant-rich film. In systems containing comparable amounts of oil and water, equilibrium bicontinuous structures in which the oil and the water domains interpenetrate in a more complicated manner are formed. In this region,

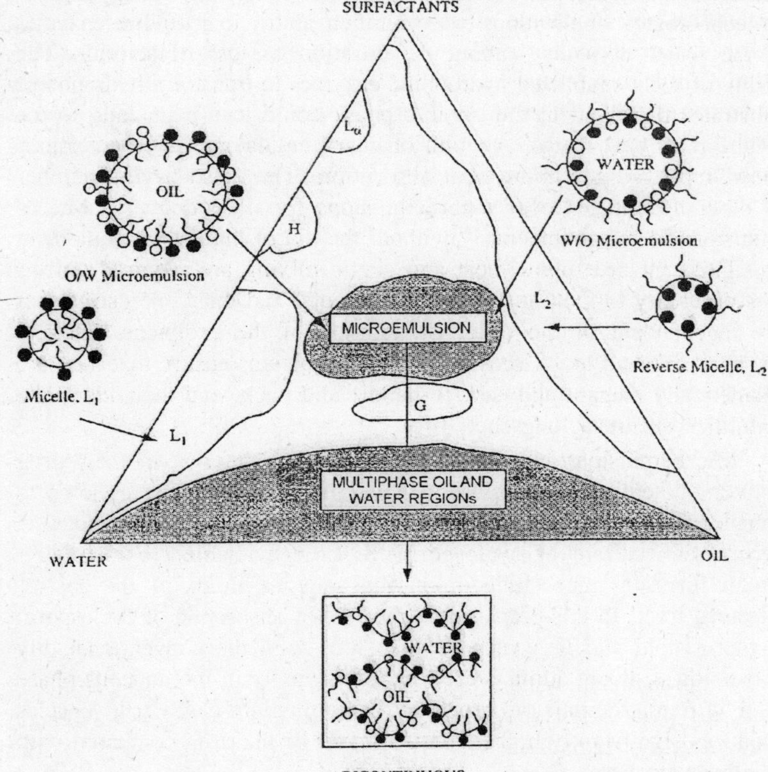

Fig. 12.4. Ternary phase diagram for oil, water, and surfactant mixtures showing micellar, microemulsion, and multiphase macroemulsion regions with schematic representations of various structures.

infinite curved channels of both the oil and the water domains extend over macroscopic distances and the surfactant forms an interface of rapidly fluctuating curvature, but in which the net curvature is near zero.

Pharmaceutical and Biological Applications of Microemulsions

Microemulsions provide ultralow interfacial tensions and large interfacial areas as well as the ability to concentrate and localize significant amounts of both oil- and water-soluble materials within the same isotropic medium. Over the years, attention has been focused on their potential use as novel reaction media for a wide range of chemical, biochemical, and photochemical reactions, and as carriers for chemicals and small particles. Inverse microemulsions of the w/o type are the subject of particular interest because of the rapidly emerging range of biotechnological applications based on their ability to solubilize enzymes in the water domains without denaturation or loss of activity. The ability of such solubilized hydrophilic enzymes to transform hydrophobic substrates dissolved in the organic phase could lead eventually to the synthesis of new drugs. As with ordinary emulsions, microemulsions show improved gastrointestinal absorption. They also have a number of other advantages over macroemulsions for drug delivery. Micro-emulsions form spontaneously without the aid of high shear equipment or significant heat input (heat and gentle mixing are required only if it is necessary to melt any of the ingredients) and their microstructures are independent of the order of addition of the excipients. Optical transparency and low viscosity of microemulsions ensure that they are cosmetically elegant and easy to handle and pack, and their indefinite stabilities ensure a long shelf life.

Microemulsions have thus attracted much interest in their drug delivery potential. Both o/w and w/o emulsions have been shown to enhance the oral bioavailability of drugs, including various peptides. A peroral concentrate of cyclosporine is now available commercially, which forms a microemulsion in the aqueous fluids of the gastro-intestinal tract. In this preparation, the rate of absorption of cyclosporin is more rapid and less variable than it is with the conventional oily dispersion. Calcein administered intraduodenally in the aqueous phase of a w/o microemulsion prepared from medium-chain triglycerides produces significantly higher plasma levels of the drug compared with an aqueous solution.

Microemulsions have also been used for topical delivery where they increase drug absorption. For example, cetyl alcohol, which is commonly used as an emulsifier in lotions and creams, is absorbed faster and deeper into the skin when formulated as a component of a microemulsion. Although efficient skin penetration may be desirable for a therapeutic agent, the relatively high concentrations of surfactant (10–25%) and cosurfactant or cosolvent (5–10%) in such formulations could enhance skin absorption of potential irritants or carcinogens. In fact, the main limitations in realizing the full potential of microemulsions as drug delivery systems are the narrow range of surfactants, cosurfactants, solvents, and other materials acceptable pharmaceutically.

Microemulsion Formation

Many approaches have been used to explore the mechanisms of microemulsion formation and stability. Early theories considered interfacial aspects of microemulsions and did not distinguish between thermodynamically stable systems and very fine kinetically stable emulsions. For microemulsions to form spontaneously, the free energy involved when the interfacial area is increased, ΔG ($\Delta G = \gamma \Delta A$, where ΔA is the increase in interfacial area) must be negative. An essential requirement is that the interfacial tension between the oil and water phases γ, is reduced to a very low value by the interfacial film, giving a small but positive free-energy value. The dispersion of the droplets in the continuous phase increases the entropy of the system. Microemulsions form because the negative free energy changes due to the entropy of the dispersion of droplets in the continuous phase overcomes the positive product of the small interfacial tension and the large interfacial area A.

The curvature of the oil–water interface in microemulsions varies from highly curved towards oil (o/w) or water (w/o) to zero mean curvature in bicontinuous structures. The type of microemulsion that forms depends on the properties of the surfactant, cosurfactant and the oil. Although there are no strict rules for choosing the appropriate microemulsion components, there are a number of general guidelines based on empirical observations. The surfactant(s) chosen for a particular oil must:

1. lower interfacial tension to a very low value to aid dispersion processes during the preparation of the microemulsion.
2. be of the appropriate hydrophile–lipophile character to provide the correct curvature at the interfacial region for the desired microemulsion type, o/w, w/o or bicontinuous.
3. provide a flexible film that can readily deform round small droplets.

The analysis of film curvature for surfactant associations leading to microemulsion formation has been rationalized by Mitchell and Ninham. They used a packing ratio *P* defined as *V/al*, where *V* is the partial molar volume of the surfactant, *a* the cross sectional area (i.e., size) of the surfactant head group, and *l* the maximum length of the surfactant chain. The packing ratio provides a direct measure of HLB and is influenced by the same factors. Oil-in-water microemulsions are favored if the effective polar part of the surfactant is more bulky than the hydrophobic part, that is, *P* varies from 0 to 1, and the interface curves spontaneously towards water (positive curvature). Water-in-oil microemulsions form when the interface curves in the opposite direction, that is, *P* is greater than 1 (negative curvature). At zero curvature, when the HLB is balanced and *P* is zero, either bicontinuous or lamellar structures may form according to the rigidity of the film. The critical packing parameter *P* is based purely on geometric considerations. Hydration of the surfactant head group and penetration of the oil and the cosurfactant into the surfactant film also affect the packing and curvature, which also illustrates how formulation variables may be manipulated to produce a microemulsion of the desired type.

Most single-chain surfactants do not lower the oil–water interfacial tension sufficiently to form micro- emulsions nor are they of the correct molecular structure, and short- to medium-chain length alcohols are necessary as cosurfactants. The cosurfactant also ensures that the interfacial film is flexible enough to deform readily around each droplet as their intercalation between the primary surfactant molecules decreases both the polar head group interactions and the hydrocarbon chain interactions. Medium-chain alcohols such as pentanol and hexanol have been used by many investigators as they are particularly effective cosurfactants. They are not, however, suitable for pharmaceuticals due to their high irritant potential. Double-chain surfactants such as anionic Aerosol-OT (bis-2-ethylhexyl sulfosuccinate) or cationic DDAB (didodecyldimethylammonium bromide), which have relatively small head groups and bulky hydrophobic portions, are already of the required HLB to form w/o microemulsions spontaneously without a cosurfactant. Unfortunately, these widely investigated surfactants are too toxic for general pharmaceutical or biotechnological applications. Double-chain phospholipids such as the phosphatidylcholines of lecithin are an obvious possibility. Although lecithin is too lipophilic to form microemulsions, pharmaceutically acceptable microemulsions have been prepared from double-chain phospholipids by using acceptable cosurfactants such as ethanol, propanol, or *n*-butanol with isopropyl myristate. Self-emulsifying drug delivery systems are composed of triglyceride oils and surfactant mixtures that undergo spontaneous emulsification when mixed with water. This principle is used in the commercial product Sandimmune Neoral, which forms a microemulsion in situ when diluted by gastric fluid.

Formulation and Preparation of Microemulsions

As microemulsions are thermodynamically stable, they can be prepared simply by blending oil, water, surfactant, and cosurfactant with mild agitation. Once the appropriate microemulsion components have been selected, quaternary phase diagrams or ternary pseudo-phase diagrams may be constructed to define the extent and nature of the microemulsion regions and the surrounding two- and three-phase domains. The microemulsion region can be identified and characterized using the range of light, neutron, and X-ray scattering and other techniques such as NMR and microscopy. Problems arise in interpretation of data in systems of high droplet volume fraction due to inter-droplet interactions. The normal practice of investigating systems at relatively low concentrations and then extrapolating to zero concentration in order to eliminate interparticle interactions cannot be applied to microemulsions as it is not possible to dilute the systems without affecting their structure. Hard sphere models, such as those adapted from Percus and Yevick, have been successfully used to analyze scattering data from concentrated w/o microemulsions.

13

Botanical Molecular Farming

Plants can be used to synthesize a wide range of industrial and pharmaceutical proteins, providing new commercial opportunities in the agriculture and biotechnology industries. Crops that were once used solely for the production of food, feed or raw materials can now produce recombinant proteins on an agricultural scale. Although plants are relative newcomers in the molecular farming marketplace, they have numerous advantages over the more traditional production systems, particularly in terms of cost, convenience, scalability and product safety. In a commercial setting, the cost of production decreases with increasing scale, and field-grown transgenic plants therefore represent the most lucrative of all the plant-based production platforms. However, controversy surrounds the biosafety of molecular farming in field plants, particularly their potential impact on human health and the environment.

Specific biosafety risks fall into two major categories, which we describe as the risk of *transgene spread* and the risk of *unintended exposure*. The risk of transgene spread can be defined as the potential for transgene DNA sequences to spread outside the intended host plants and production site. This can result in the growth of transgenic crops in fields reserved for non-transgenics, the growth of transgenic crops in non-cultivated areas, the spread of foreign DNA to other plants (and possibly to microbes and animals) and the uncontrolled production of recombinant proteins in natural settings. Mechanisms of transgene spread include the dispersal of transgenic plants or seeds by human and animal activities or the weather, outcrossing via transgenic pollen, and horizontal gene transfer from plants to other organisms. The risk of unintended exposure can be defined as the potential for any non-target organism (including humans) to come into contact with the recombinant protein produced by a transgenic plant. Many different mechanisms can be involved, including herbivory and parasitism, the exposure of pollinating insects to transgenic pollen, the exposure of microbes in the rhizosphere to root exudates, the exposure of non-target microbes and animals to proteins secreted in the leaf guttation fluid, the release of recombinant proteins by dead and decaying transgenic plant material, and the contamination of food or feed crops during harvesting, transport, processing and/or waste disposal. In many cases, transgene spread can lead to unintended exposure because the naturalization of transgenic plants outside the intended production site results in the wider exposure of non-target organisms. While these risks apply to all field-grown transgenic crops regardless of their use, those used for molecular farming deserve special attention because of the pharmacological or toxic properties of many of the recombinant proteins they produce. A final reason for concern, at least to the biotechnology industry, is that the spread of proprietary transgenes into wild species places intellectual property in the public domain. In this chapter, we discuss the biosafety issues associated with molecular farming and some of the emerging strategies that are being used to address them.

Transgene Spread

Classes of Foreign DNA Sequences in Transgenic Plants

Three different classes of DNA sequence need to be considered when addressing the biosafety aspects of molecular farming. The first class can be described as the *primary transgenes*, i.e. the genes and surrounding elements required to express the desired recombinant product. Note that the term *transgene* has a much broader meaning than the word *gene*, from which it is derived, and generally refers to a DNA cassette that may include one or more actual genes plus any regulatory elements and other sequences needed for proper expression. Primary transgenes are absolutely required in molecular farming since without them there would be no production of the desired protein. The impact of such sequences on the survival and fecundity of wild species is difficult to predict but it is certainly undesirable for proteins that have potent pharmacological or immunological effects when administered to humans or animals to be expressed in natural populations of plants and microorganisms, or in crops intended for the human and domestic animal consumption. The second class of sequences can be described as the *secondary transgenes*, i. e. the genes and surrounding elements that are needed during transformation and regeneration but which are not essential for continued production of the target recombinant protein. This group includes selectable marker genes, reporter genes and genes encoding other accessory proteins that are used to manipulate primary transgenes or their expression (e.g. recombinases), and the regulatory elements required for their expression. These sequences need to be introduced during the gene transfer process but can be discarded when stable plant lines are available. The impact of secondary transgenes on the survival and fecundity of wild species is also difficult to evaluate but there is a great deal of concern that certain markers could have negative effects if they spread outside the intended transgenic plants.

In particular, there is concern that herbicide-resistance markers could spread to weedy plants, producing a new generation of 'superweeds', and that antibiotic resistance markers could spread to pathogenic bacteria, severely compromising the use of antibiotics in human healthcare. The final category of sequences can be described as *superfluous DNA*, and comprises those sequences that are required neither for transformation nor for recombinant protein synthesis, but which tend to be introduced during the transformation process. Essentially, this means vector backbone sequences from plasmid vectors, which are linked to the primary and secondary transgenes.

Mechanisms of Transgene Pollution – Vertical Gene Transfer

Vertical gene transfer is the movement of DNA between plants that are at least partially sexually compatible. This is the most prevalent mechanism of transgene spread and occurs predominantly via the dispersal of transgenic pollen, resulting in the formation of hybrid seeds with a transgenic male parent. Gene flow from transgenic to non-transgenic populations of the same crop occurs by this method if the two populations are close enough for wind- or insect-mediated pollen transfer. Very high rates of gene flow from crops to related wild species have also been documented along this route. For example, Kling noted that 50% of wild strawberries growing near a field of cultivated transgenic strawberries contained marker genes from the transgenic population. Similarly, herbicide resistance genes have introgressed from transgenic oilseed rape (*Brassica napus*) into its weedy cousin *B. campestris* by hybridization. As discussed below, a number of potential solutions to the problem of transgene pollution have been based on preventing the spread of transgenic pollen, either by physical or genetic containment. However, hybrid seeds can also be generated with the transgenic plant as the female parent if the transgenic crops are fertilized by wild type pollen. In this case, transgene pollution would occur via seed dispersal, either during growth, harvesting or during transport. Seed dispersal from fully transgenic plants can also result in the colonization of natural ecosystems and is more prevalent if seeds can lie dormant for extended periods.

Mechanisms of Transgene Pollution – Horizontal Gene Transfer

Horizontal gene transfer is the movement of genes between species that are not sexually compatible and may belong to very different taxonomic groups. The process is common in bacteria, resulting in the transfer of plasmid-borne antibiotic resistance traits from harmless species or strains to pathogenic ones, but there are few examples of natural gene transfer between bacteria and higher eukaryotes. *Agrobacterium* spp. represent a special case where gene transfer occurs naturally from bacteria to plants if the bacterium contains an appropriate virulence plasmid. There is a perceived risk that horizontal gene transfer from transgenic plants to bacteria in the soil or in the digestive systems of animals could yield new bacterial strains expressing primary and/or secondary transgenes. These traits could have unpredictable effects on relationships between different organisms, e.g. they could render harmless bacteria pathogenic, or could be passed on to pathogenic species making them more difficult to control. There is a specific concern that antibiotic resistance markers and transgenes encoding pharmaceutical proteins could be acquired by human pathogens.

The risks of horizontal transgene transfer from plants to microbes are considered to be extremely small because of the lack of evidence, over millions of years of evolution, that natural plant genes have followed this route. For example, Kay *et al*. demonstrated horizontal transfer of marker genes from the chloroplasts of transplastomic tobacco plants to opportunistic strains of *Acinetobacter* spp., but transfer was achieved only under highly idealized conditions in which the bacteria were modified to contain a sequence homologous to the plant's transgene. No gene flow was demonstrated to wild type strains of the bacterium. Even if gene transfer from plants to bacteria did occur in nature, it would be necessary for the transgene to be maintained in the recipient bacterial population. In the case of antibiotic resistance markers there might be strong selective pressure for transgene maintenance due to the widespread use of antibiotics. However, since all natural plants are already liberally covered with antibiotic-resistant bacteria, these would appear to be a much more likely source of resistance genes that could jump to human pathogens. DNA can be taken up from saliva by oral bacteria, and cells lining the gastrointestinal tract can take up and incorporate DNA from the gut. Again however, there is a conspicuous lack of evidence that such mechanisms have resulted in the stable incorporation of a plant gene into a bacterial population. Studies with glyphosate-resistant transgenic plants showed that the DNA was completely digested in the gastric environment within a few minutes. Antibiotic resistance genes are the focus of attention because of their potentially strong and general selective advantage in human pathogens. Other transgenes, with much more specific therapeutic applications, would not provide the same benefits as antibiotic resistance and would likely be eliminated even if transfer from plants to bacteria were inevitable. These seemingly insurmountable barriers indicate that horizontal gene transfer is unlikely to represent a significant hazard, and biosafety research has therefore focused on ways to prevent transgene spread by vertical gene transfer.

Combating the Vertical Spread of Transgenes

Choosing an Appropriate Host

An appropriate choice of host species can go a long way to prevent or minimize transgene spread by dispersal or vertical gene transfer. In general terms, plants that produce large amounts of pollen or large numbers of seeds should be avoided, especially if the seeds are small and easily dispersed. Plants that are often grown as open-pollinated varieties or those that cross spontaneously with wild relatives are also to be avoided, while self-fertilizing plants would be a better choice. Certain plants have been singled out as inappropriate hosts by regulatory organizations such as APHIS. For example, alfalfa and canola have been highlighted as unsuitable because they are bee-pollinated, sexually compatible with abundant and local weed species and the seeds can lie dormant for several years, making volunteer plants difficult to isolate and destroy. In the end, however, the search for the ideal crop in terms of biosafety will often frustrate the very principles upon which molecular farming in plants is based, i. e. large-scale production, rapid scale-up due to prolific seed production, and the use of existing agricultural and processing infrastructure. There is no single field crop that meets all biosafety demands, and further steps in addition to the selection of a host species must therefore be taken to limit outcrossing and other forms of vertical gene transfer.

Using only Essential Genetic Information

One way in which the risk of transgene spread can be minimized is to limit the amount of new genetic material incorporated into the production crop. As discussed above, while only the transgene encoding the recombinant protein is required for protein production, transformation usually involves a host of other sequences including superfluous backbone elements and selectable markers. The standard method for producing a transgenic plant line is to introduce the primary transgene along with a selectable marker, which allows the propagation of transformed plant material at the expense of non-transformed material. The use of selectable markers is perhaps one of the major issues in biosafety because traditional markers, which exploit herbicide or antibiotic resistance as selectable traits, are each thought to represent significant environmental or health threats. It is also standard practice to transform plants with plasmid vectors containing the expression cassette. This results in the integration of vector backbone sequences along with the functional primary and secondary transgenes. Not only are such sequences superfluous to requirements, but they also have numerous undesirable effects in transgenic plants, acting as triggers for de novo methylation and promoting extensive rearrangement of the foreign DNA sequences prior to integration. They may also carry additional functional DNA sequences such as selectable markers, promoters and origins of replication used in bacteria, which could become active after gene transfer to non-target organisms.

Ideally, it would be possible to produce transgenic plants carrying just the primary transgene, without recourse to marker genes and other superfluous sequences. The negative impact of these sequences has been established only in the last few years, and only recently have efforts been made to dispense with them. In the case of *Agrobacterium*-mediated transformation, it has been realized that inefficiency in the T-DNA processing step results in the co-transfer of vector sequences in 30–60% of transformation events depending on plant species, *Agrobacterium* strain and transformation method. Since plasmids are pre-requisite for this mode of gene transfer, the only way to guarantee clean transformation (transformation without vector sequences) is to flank the T-DNA with counterselectable marker genes that kill any plant cells containing them. With direct DNA transfer methods (such as PEG-mediated protoplast transformation, electroporation and particle bombardment), vector sequences are generally present in all transformants because whole plasmids are used in the transformation procedure. An efficient and practical alternative is to carry out transformation using minimal cassettes, i.e. linear constructs containing just the promoter, open reading frame and polyadenylation signal. Not only does this avoid vector backbone integration but it appears to circumvent another problem specific to direct DNA transfer methods, which is the formation of large, highly complex, multicopy transgene loci containing many rearrangements. Such loci are undesirable because they tend to be unstable, and in many cases contain inverted repeats or truncated transgenes that have the potential to form DNA secondary structures or to express hairpin RNAs, both of which can trigger transgene silencing. In contrast, transformation with minimal cassettes leads to the generation of very simple integration patterns with the majority of transgenic loci represented by a single transgene copy.

Dispensing with selectable markers is more difficult because stable transformation is a rare event and markers are required to identify the very few transformed plant cells in a large background of nontransformed ones. It is possible, although quite laborious, to screen plant cells for the incorporation of a primary transgene using the polymerase chain reaction, without relying on any type of marker. However, most '*marker-free*' transformation strategies involve removal of selectable markers *after* transformation has been achieved. An alternative approach is to use an innocuous scorable marker gene such as *gus*A (encoding the bacterial enzyme-glucuronidase) or *gfp* (encoding the jellyfish green fluorescent protein). Even better, a bacterial gene or preferably a plant gene can be used as an innocuous selectable marker, i.e. a gene that would have no conceivable negative effects in wild populations. Examples of such markers include growth regulators (e.g. *ipt* or *CKI1*) and metabolic markers (e.g. *manA* or *BADH*) under inducible control. Such markers could be used to restrict the growth of plants under non-permissive conditions but would not affect the growth or reproduction of wild plants.

Elimination of Markers After Transformation

Where the use of conventional markers is inescapable, an acceptable strategy is the elimination of these genes after transformation, leaving transgenic plants containing the primary transgene alone. This can be achieved either by segregation or recombination, the former requiring independent cointegration of the marker and primary transgene and the latter requiring the use of site-specific recombination systems such as Cre-*lox*P or FLP-*FRP*.

It is surprisingly difficult to persuade separate transgenes to integrate at different loci allowing segregation in later generations.

Where two separate plasmids are used to coat microprojectiles, cointegration at the same locus is the predominant outcome (usually as a highly complex concatemer). The introduction of separate binary vectors into *Agrobacterium tumefaciens*, and even the use of different *A. tumefaciens* strains for co-infection, also generally results in co-integration, although this depends on the strain. For example, Komari *et al.* were able to achieve marker gene segregation in a small number of R1 transgenic plants following a transformation strategy involving co-infection with two different *A. tumefaciens* strains. More recently, it has been shown that particle bombardment with minimal cassettes can yield a large number of independent cointegration events, resulting in efficient marker gene segregation in later generations. An alternative and rather elegant way to achieve the same goal is to clone the primary transgene and marker gene in a single construct, but enclose the marker gene within the active elements of a transposon such as *Activator*. Integration is followed by transposition, resulting in the relocation of the marker gene to a different genomic site. As discussed above, the marker can then be eliminated by crossing.

The need for crossing can be avoided by building a marker excision strategy into the transformation construct. In most cases, this involves the use of a two-component site-specific recombination system such as Cre-*loxP*. Cre is a recombinase that recognizes short sequences known as *lox*P. If two *lox*P sites are in the same orientation, Cre recombinase activity will excise any DNA between them, so marker genes flanked by *lox*P sites can be efficiently excised from transgenic plants if Cre is present. Cre can be expressed transiently or crosses can be carried out between primary transgenic lines and Cre-transgenic lines to generate hybrids containing both *cre* and the *lox*P-flanked marker, allowing the marker gene to be removed. Where this strategy is used, further crossing may be required to remove the *cre* transgene, unless a '*self-excising*' *cre* transgene is integrated. More recently, the *att*B system from bacteriophage λ has been developed for use in transgenic plants because spontaneous recombination occurs at a high frequency, leading to marker removal.

Site-specific recombination has also been used to reduce the complexity of multicopy transgenic loci generated by particle bombardment. As discussed above, such loci are prone to transgene silencing and structural instability, and are unsatisfactory from a biosafety perspective because the complex organization means that uncharacterized transcripts and proteins could be produced with unpredictable effects. Simplification is possible either by inserting the transgene at a predefined locus or by streamlining the locus structure after transformation. Both these processes can be achieved using a site-specific recombination system such as Cre-*lox*P. Site-specific integration of transgenes can occur if the genome contains a recombinase recognition site such as *lox*P that has been introduced in a previous round of transformation. Transgene integration occurs at a low efficiency if an unmodified recombination system is used because the equilibrium of the reaction favors excision. However, high-efficiency Cre-mediated integration has been achieved in tobacco using mutated *lox*P sites. Post-integration locus simplification in transgenic wheat has been achieved by incorporating a single *lox*P site within the transgene. Cre expression then drove recombination between the tandemly-arranged *lox*P sites until only one site remained, reducing the transgenic locus to a single copy. This resulted in increased transgene expression accompanied by reduced methylation at the transgenic locus.

Containment of Essential Transgenes

For indispensable primary transgenes, the only way to avoid transgene spread from field plants to compatible crops and wild species is by containment. The aim of containment is to prevent seed and pollen dispersal, prevent the survival of dispersed seeds and pollen, or prevent gene flow from viable pollen. The containment may be physical and based on habitat barriers. For example, transgenic plants can be maintained in greenhouses, in artificially-irrigated desert plots miles from any other plants, or in underground caverns and caves. Alternatively, the physical containment may be focused on individual plants. For example, flowers can be emasculated before viable pollen has developed, or the flowers/fruits may be concealed in plastic bags. Isolation zones are often placed around transgenic crops. These can be barren, but a more suitable alternative for insect-pollinated crops is to provide a zone of non-insect-pollinated plants which would discourage the insects from leaving the transgenic zone. Barrier crops, i.e. a border of non-GM plants of the same species as the transgenic crop, are also useful as these can absorb much of the pollen released by transgenic plants and can then be destroyed after flowering.

Biological containment measures provide additional barriers to gene flow and many different strategies have been tested. In some cases, natural genetic barriers have been exploited. For example, molecular farming in self-pollinating species (e.g. rice, wheat, pea) or crops with no sexually compatible wild relatives near the site of production provide a first level of defense against gene flow. Similarly, crops with asynchronous flowering times or atypical growing seasons are useful. Cleistogamy (self-fertilization before flower opening) is an extension of the above, and could be engineered into crops used for molecular farming by modifying the architecture of flower development. In practice, however, there is always a residual risk of outcrossing. Another potential strategy, yet to be fully explored, is the exploitation of apomixis (embryo development in the absence of fertilization). Transformation strategies can also be adapted to take advantage of natural barriers. An example of this approach is genomic incompatibility, which is suitable for polyploid species such as wheat. Many cultivated crops are polyploid but have distinct genomes, only a subset of which are compatible with related wild species for interspecific hybridization. In the case of wheat, only the D genome is compatible with wild *Aegilops* species. Therefore, wheat plants used for molecular farming should carry the transgene(s) on the A or B genomes, a fact that can be established by fluorescence in situ hybridization (FISH) before the plants are transferred to the field.

These natural mechanisms may be replaced or augmented with artificial genetic strategies which are themselves controlled by transgenesis. Such strategies include male sterility, chloroplast transformation, conditional transgene excision and transgene mitigation. Male sterility is achieved by interfering with flower development, or more specifically pollen development, often through the expression of a ribonuclease that prevents the differentiation of the male reproductive organs. For example, Bayer Crop Sciences have developed and commercialized a male sterile variety of oilseed rape expressing barnase. The barnase inhibitor (barstar) is also expressed, but it is controlled by an inducible promoter allowing propagation of the transgenic line under laboratory conditions but not in the field. This strategy prevents outcrossing by pollen dispersal but not by pollen immigration, so there remains the possibility of transgene pollution by seed dispersal.

An alternative to male sterility is chloroplast transformation, i.e. the introduction of foreign DNA into the chloroplast genome rather than the nuclear genome. This limits gene flow because the pollen of many crop species does not contain chloroplasts, and where chloroplasts are present, functional DNA is either not transferred to the egg during fertilization, or is degraded during generative and sperm cell development. There are several advantages to molecular farming by chloroplast expression in addition to the biosafety benefits, including the high transgene copy numbers in photosynthetic cells, the absence of position effects and transgene silencing phenomena which can result in low yields in nuclear transgenic plants, and the opportunity to carry out multigene engineering using operons. Thus far, the technology is only applicable to three field species used for molecular farming: tobacco, tomato and potato. However, transformed chloroplasts in these species have been used successfully for the production of diverse products, including biopolymers, vaccines and human growth hormone. One possible disadvantage is that proteins produced in chloroplasts are not glycosylated so this system cannot be used for the production of complex glycoproteins. It is also notable that chloroplast inheritance is not strictly maternal in some species, so while gene flow by pollen dispersal may be limited, it may not be eliminated. As with male sterility, chloroplast transformation does not prevent transgene spread by volunteer seed dispersal.

Another genetic barrier to transgene flow is seed sterility, which is achieved by using suicide genes to destroy the developing plant embryo. This mechanism is employed in Monsanto's notorious '*terminator technology*' in which a ribosome inhibitor protein is expressed under the control of an embryonic promoter, but in a manner that is regulated by tetracycline. Among several variations of the technique originally discussed in a patent application assigned to Pine Land Corporation, one involved the tetracycline-depended expression of Cre recombinase, which would lead to the excision of the suicide gene if it was flanked by *lox*P sites. A 'recoverable block of function' system, based on the constitutive expression of barnase and the inducible expression of barstar, has also been developed.

Transgenic mitigation can also be used to prevent the spread of transgenes to wild plants. This involves the inclusion of a tightly linked transgene that confers a trait that is selectively neutral to the crop

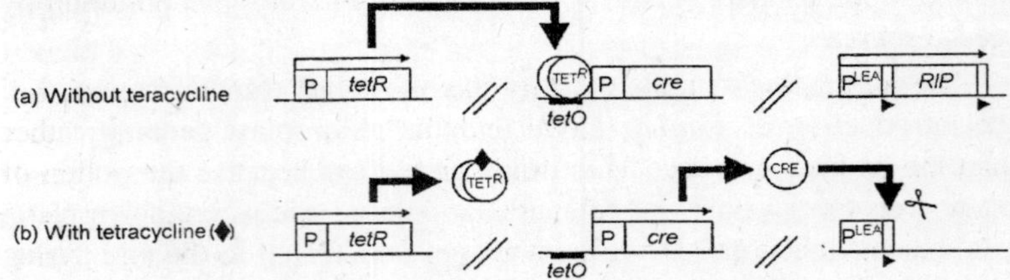

Fig. 13.1. The 'terminator technology' which can be used to prevent the growth of volunteer plants from dispersed transgenic seed.

but disadvantageous to wild plants. Examples might include dwarfing genes or genes that control seed dormancy or shattering. A more sophisticated strategy is conditional transgene excision.

In this strategy, plants are created with the transgene flanked by *lox*P sites. A *cre* transgene is also present, and this is expressed under the control of a cell-specific or inducible promoter, such that the transgene is physically removed before flowering. If the *cre* transgene is also present within the *lox*P sites, then this transgene will be removed from the plant at the same time, but only when its 'clean-up' task is complete. One potential drawback of this approach is that incomplete transgene excision will leave a residual population of transformed cells from which transgenic gametes could arise.

Unintended Exposure to Recombinant Proteins

Environmental Risks of Unintended Exposure

The recombinant proteins produced by transgenic plants constitute another risk to the environment. One immediate concern is the possible negative effect of recombinant proteins on non-target organisms, particularly insects and microorganisms that interact directly with the plant and herbivores that may eat transgenic plant material laced with industrial enzymes or protein drugs. Such proteins might have direct toxicity effects, or they might accumulate in the food chain and therefore affect animals that do not interact with the transgenic plants at all. Toxicity may result from direct consumption (e.g. the ingestion of toxins by aphids, and knock-on effects to lady-bugs and birds further up the food chain), by simple exposure to the plant (e.g. the effects of pollen on butterflies and moths), from the exudation of recombinant protein into the rhizosphere or leaf guttation fluid (most likely to affect microorganisms) and by the consumption of dead and decaying plant material by saprophytes.

Many recombinant proteins expressed in plants are directed to the secretory pathway in order to fold or assemble properly. Such proteins accumulate in the apoplast, the space beneath the cell wall, but there is some leakage into the guttation fluid and root exudate that may change the biochemical environment on the exposed leaf surface or in the rhizosphere. The long term effects of recombinant pharmaceuticals accumulating in the soil and in drainage water have not been investigated and provide scope for all manner of unseen hazards. Finally, the processing of transgenic plants will produce waste containing residual recombinant proteins. An important biosafety issue, particularly for large-scale molecular farming enterprises, is what to do with this waste plant material. A pertinent danger is that such material will be allowed to decay in the environment, providing further opportunities for both protein pollution and transgene escape.

Like transgene spread, the risk of unintended exposure to recombinant proteins can be addressed to a certain extent by physical containment, since this restricts the impact of protein toxicity to a very localized environment. In other words, although microbes in the rhizosphere of contained transgenic plants are exposed to the same extent as those associated with uncontained transgenic plants, the microbes themselves are contained, thus limiting knock-on effects to non-target organisms. However, further barriers to protein toxicity can be put in place by controlling transgene expression or protein structure, therefore limiting the availability of the protein even to closely interacting organisms.

Addressing the Risks of Unintended Exposure

Controlling transgene expression

The exposure of non-target organisms to recombinant proteins can be minimized by restricting expression to particular tissues. For example, a number of promoters have been identified that restrict gene expression to seeds, tubers or fruit. This prevents the consumption of recombinant proteins by insects and other animals feeding on green plant tissue, and likewise prevents other forms of contact, such as the exposure of pollinating insects to recombinant proteins expressed in pollen grains. By avoiding transgene expression in roots, leaching of the recombinant protein into the soil (and consequent disruption of the rhizosphere) is also prevented. If restricted expression strategies are used in combination with effective management (e.g. specific harvesting times) then vegetative transgenic material can decay safely in the environment with little risk of protein contamination in the environment or unintended exposure.

In monocots, where seed expression is the normal strategy, various seed-specific promoters usually derived from seed storage protein genes have been employed to control transgene expression. Examples include promoters from maize zein, rice glutelin and pea legumin genes. Care must be taken, however, because although these promoters are described as seed-specific, a low level of activity is present in other tissues. One pertinent example is the bean USP (unknown seed protein) promoter, which has been used in transgenic peas. Although predominantly seed-specific, this promoter also drives low level transgene expression in pollen grains, which could pose a risk to pollinating insects if it were to be used in insect-pollinated plants. An alternative strategy is to bring the transgene under inducible control, such that the recombinant protein would be expressed only when the plant was exposed to a certain chemical inducer. One of the most promising developments in this area is the use of inducible expression systems to prevent recombinant protein expression until after the crop has been harvested, as has been shown for recombinant glucocerebrosidase using a tomato promoter induced by mechanical stress. A more recent example is the peroxidase gene promoter from sweet potato (*Ipomoea batatas*), which is induced by hydrogen peroxide, wounding or ultraviolet light. In all cases, an effective waste-management policy is necessary to dispose of the waste generated by product processing and extraction.

Controlling protein accumulation and activity

In addition to the control of transgene expression, the protein can also be targeted to a specific intracellular compartment. This would not necessarily protect herbivores from exposure to the protein, but it might limit adventitious contact. For example, by adding a KDEL tetrapeptide tag to the C-terminus of a recombinant protein which has been targeted to the secretory pathway with a suitable N-terminal signal sequence, there is efficient retrieval from the Golgi apparatus to the *endoplasmic reticulum* (ER). This helps to prevent proteins being secreted to the apoplast, phloem or xylem, where contact with the plant's environment, including microbes and insects, becomes more likely. This is also the case for recombinant proteins containing a heterologous trans-membrane domain, which are anchored in the plasma membrane or in the vacuolar membrane depending on other targeting information.

The chloroplast or vacuole are alternative destinations for recombinant proteins produced in plants, and help to protect the plant from toxicity effects as well as preventing unintended exposure. Recombinant proteins can also be produced as inactive precursors that have to be processed by proteolytic cleavage before they attain full biological activity. This strategy has been used by Prodigene Inc. for the production of proteases such as trypsin and was also used for the expression leech hirudin. As is the case for targeting to the chloroplast and vacuole, the expression of inactive precursors not only limits the extent of protein toxicity in the environment, but also protects the host plant from any negative effects the recombinant protein might have on growth or development.

Contamination of the food chain during processing

We have discussed ways in which transgenic plant material, or products derived therefrom, could enter the food chain of humans or domestic animals. These include transgene spread to food and feed crops, contact between transgenic plants and non-target organisms, adventitious herbivory of transgenic plants and leaching of recombinant proteins into the environment through poor waste management. Another major source of contamination is the unintentional mixing of transgenic and non-transgenic crops during harvesting, transport, refining and processing, which has resulted in some highly-publicized incidents including the discovery of recombinant DNA in Linda McCartney food products and the discovery of unregistered Star-link corn in maize products. A more pertinent example in molecular farming is the recent ProdiGene incident in which stray maize plants expressing pharmaceutical proteins were found growing among a soybean crop.

The problem of contamination is compounded by the use of existing facilities to process both food/feed crops and crops used for molecular farming. Ideally, there should be a clear distinction between transgenic plant material used for molecular farming and any normal plant material being processed in the same facility, which is intended for human or domestic animal consumption. A rigorous series of regulatory practices should be in place from the farm to the factory, ensuring complete isolation of transgenic material during growth, harvesting, transport, storage, processing, extraction and waste disposal, and this should supported by validated procedures for cleaning shared equipment. The accidental mixing of transgenic and non-transgenic harvest products is more likely when those products appear visually identical. Therefore, an important step towards identity preservation is the use of noncommercial crop varieties, visually striking varieties (e. g. white tomatoes) or non-food/feed crops that could not possibly be introduced into food or feed processing by misidentification (e.g. tobacco).

Molecular farming in field plants provides an opportunity for the economical and large-scale production of pharmaceuticals, industrial enzymes and technical proteins that are currently produced at great expense and in small quantities. However, this opportunity is not risk free, and measures for environmental protection must be put into place to make sure that the benefits of molecular farming are not outweighed by risks to human health and the environment. In this chapter, we have summarized the available strategies that can be used to limit the amount of unnecessary foreign DNA incorporated into transgenic plants, prevent transgene spread to non-production plants and other organisms, and limit the exposure of non-target organisms, including humans, to the recombinant products synthesized in plants. The production of well-characterized transgenic plants will allow more effective risk assessment and transgene tracking, and combinations of management and containment strategies will help to prevent transgene spread, protein toxicity and contamination of the food and feed chains. Whatever precautions are taken, it is unlikely that these undesirable occurrences will be completely eliminated so it is possible that the benefits of plant-based protein synthesis will be exploited less controversially in highly contained bioreactors using aquatic plants, single celled plants, or plant cell suspension cultures, albeit with the loss of many of the economical and scalability advantages of field crops.

14

Extrusion and Extruders

Extrusion is the process of forming a raw material into a product of uniform shape and density by forcing it through an orifice or die under controlled conditions. An extruder consists of two distinct parts: a delivery system which transports the material and sometimes imparts a degree of distributive mixing, and a die system which forms the material into the required shape. Extrusion may be broadly classified into molten systems under temperature control or semisolid viscous systems. In molten extrusion, heat is applied to the material in order to control its viscosity to enable it to flow through the die. Semisolid systems are multiphase concentrated dispersions containing a high proportion of solids mixed with a liquid phase. Extrusion is achieved by formulation to control the viscosity of the semisolid mass.

Extrusion is a continuous process that affords a consistent product at high throughput rates. The process has diverse applications in a range of industries utilizing extrusion equipment specifically designed or adapted to form a particular product. A description of the different types of extruders is given here, along with details that illustrate the versatility of extrusion processing.

Theory and Characterization of Extrusion

The various types of extruders have the common feature of forcing the extrudate from a wide cross section through the restriction of the die. The force required and the characteristics of the extrudate produced are dependent on the rheological properties of the extrudate, the design of the die, and the rate at which the material is forced through the die. The theoretical approach to understanding the systems, therefore, is generally associated with dividing the process of flow into three sections: (1) entry into the die, (2) flow through the die, and (3) exit from the die.

Extrusion is dependent on the material, and the technique varies with the material studied. With regard to pharmaceuticals, most systems consist of particles dispersed in a fluid and, although consideration will be given to plastics, the main emphasis is on paste extrusions. These differ in the fact that a fluid is present between solid particles. The relative position of solid and liquid can change during the various stages of the extrusion process, and hence produce effects different from those associated with single-phase systems.

If the die is considered as a simple capillary flow, the relationship between the rate of shear (γ) and die wall shear stress (τ_w) can be described by Eq. (1):

$$\tau_w = \gamma P \cdot R/2L \qquad \ldots(1)$$

where P is the pressure drop across the length of capillary L and radius R. Corrections for entrance effects modify this equation by considering an increase in the length of the capillary to give Eq. (2):

$$\tau_w = \gamma P/2(L/R + n_b) \qquad \ldots(2)$$

where n_b is the Bagley entrance correction. Determination of n_b can be made by measuring the pressure necessary to force extrudate through dies of different length-to-radius (L/R) ratio. Extrapolation of the graphs to zero pressure values gives the value of n_b as the intercept on the L/R axis.

Han and Charles found experimentally that the exit pressure is actually above atmospheric pressure and proposed modification of Eq. (2) to Eq. (3) corrected for exit pressure losses:

$$\tau_w = (\gamma P - P_e)/2(L/R + n_b) \qquad \ldots(3)$$

where P_e is the exit pressure. This value is difficult to determine and, because it is considerably lower than the pressure loss upstream and through the die, it is usually neglected. The upstream pressure loss can be considerable and can be determined as the intercept on the pressure axis at zero L/R ratio (the Bagley equation), giving Eqs. (4)–(5):

$$\tau_w = (P_T - P_0)R/2L \qquad \ldots(4)$$

$$P_T = P_0 + 2\tau_w(L/R + n_b) \qquad \ldots(5)$$

The upstream pressure loss includes pressure losses due to kinetic energy, head effects, elastic losses, and turbulence. Harrison found for a series of pharmaceutical systems that the value of P_0 increases with increasing rate of passage through the die.

Rheological Curves

In addition to determination of the upstream pressure loss P_0 and the end correction n_b, Eq. (5) can be seen to provide a value for the shear stress τ_w in such systems. The slope of the graph P_T vs. L/R has a gradient of $2\tau_w$; that is, the die-wall shear stress. The rate of shear at the die wall—$(d\nu/dr)_w$—can be derived from the Hagen–Poiseuille's law, as in Eq. (6):

$$(d\nu/dr)_w = 4Q/R^3 \qquad ...(6)$$

where Q is the volumetric flow rate and R the radius of the die. This assumes that the flow is Newtonian. If this is not the case, Jastrzebski suggested that a correction should be made for the rate of shear, as in Eq. (7):

$$-(d\nu / dr)_w = \left(\frac{3n' + 1}{n'}\right)\frac{Q}{\pi R^3} \qquad ...(7)$$

where n' is the degree of non-Newtonian flow; it is determined from the gradient of the graph of log-shear stress as a function of the log apparent shear rate. Wilkinson has also indicated that these equations assume that: (i) the flow is laminar, (ii) there is no slip at the die wall, and (iii) the rate of shear depends only on the shear stress at the point of measurement and is independent of time.

Determination of shear rate vs. shear stress curves by application of the ram extruder allow characterization of the rheological properties of the extruded material according to the basic type of curve, as expressed by Eqs. (8)–(11).

Newtonian:

$$\sigma_w = \gamma' \eta \qquad ...(8)$$

where η is the apparent viscosity and γ' is the rate of shear.

Bingham body:

$$\tau_w = \sigma_y + \gamma' U \qquad ...(9)$$

where σ_y is the stress necessary to be exceeded before Newtonian flow commences, yield value, and U is the plastic viscosity.

Power-law model:

$$\tau_w = K\gamma^{n'} \qquad ...(10)$$

where K is the power-law viscosity constant and n' is the degree of non-Newtonian flow. For values of n' less than 1, the material becomes less viscous with increasing shear rate (shear thinning), and for values of n' greater than 1, the viscosity increases with increasing shear rate (shear thickening).

Herschel–Buckley model:

$$\tau_w = \tau_y + K\gamma^n \qquad ...(11)$$

which allows for a system that has a yield value and a shear rate dependent on viscosity.

The application of these types of flow curves requires homogeneous materials that do not change in consistency with extrusion. Harrison, Newton, and Rowe, found this not to be the case, and suggested that this was due to the presence of plug flow within the extrudate bulk and slip flow at the die wall.

The inability of the standard rheological models to quantitatively describe the process of flow into, through, and out of the die requires an alternative treatment. From a study of ceramic catalyst pastes, Benbow and Ovensten and Benbow assumed that there was broad plug flow at the center of the extrudate, with shearing occurring within a thin liquid layer at the die wall. Assuming that this layer behaves as a Newtonian liquid of thickness x and viscosity η, and that the initial shear stress to induce flow is τ_0, the total die-wall shear stress τ_w at a given extrudate velocity V is given by Eq. 12:

$$\tau_w = \tau_0 + (\eta/x)V \qquad ...(12)$$

The values of η and x cannot be determined directly and, therefore, Benbow, Oxley, and Bridgwater, introduced the term β, the die land viscosity factor, to replace η/x, as in Eq. (13):

$$\tau_w = \tau_0 = \beta V \qquad ...(13)$$

Incorporation of this expression into Eq. (4) yields:

$$P_\tau = P_0 + 2(L/R)(\tau_0 + \beta V) \qquad ...(14)$$

The value of τ_0 can be determined by plotting the extrusion pressure against the extrudate velocity V for extrusion through dies of constant value of L. The extrapolated value of extrusion pressure at $V = 0$ gives the value P_{0v0} at zero velocity, as shown by Eq. (15):

$$P_{\tau v0} = P_{0v0} + 2(L/R)\tau_0 \qquad ...(15)$$

Thus, a graph of $P_{\tau v0}$ vs. L/R provides the value of τ_0 as equal to half the slope. The value of β can be calculated by rearranging Eq. (14) to Eq. (16):

$$\beta = (P_{\tau w} - P_{0w}) - (P_{\tau v0} - P_{0v0})2(L/R)V \qquad ...(16)$$

where $P_{\tau w}$ is the total extrusion pressure at extrudate velocity V, and P_{0w} is the upstream pressure loss at extrudate velocity V.

Further characterization of the system was suggested by Benbow and Benbow and Bridgwater in terms of a yield value σ_y associated with the convergence of flow from the wide cross section of the feed to the narrow cross section of the die. This takes the form of Eq. (17):

$$P_0 = \sigma_y \ln(A_0/A) \quad ...(17)$$

where A_0 is the initial cross-sectional area and A that of the die. If the original and final cross sections are circular, Eq. (18) holds:

$$P_0 = 2\sigma_y \ln(D_0/D) \quad ...(18)$$

where D_0 and D are the barrel and die diameters, respectively. For materials that deform plastically and are time independent, the value of σ_y can be calculated from the intercept of the pressure axis divided by twice the natural log reduction ratio (D_0/D) for plots of P against L/R.

By combining this concept with those expressed above, Benbow, Oxley, and Bridgwater, and Benbow and Bridgwater further modified the Bagley equation to Eq. (19):

$$P_T = 2(\sigma_{y0} + \alpha V) \ln(D_0/D) + 2(L/R)(\tau_0 + \beta V) \quad ...(19)$$

If the die land velocity factor β varies with the extrusion rate or the liquid layer at the die wall is non-Newtonian, Eq. (19) must be further modified to Eq. (20):

$$P_\tau = 2\sigma_y \ln(D_0/D) + 2(L/R)\ (\tau_0 + \beta^* V^{1-n}) \quad ...(20)$$

where β^* is a modified power-law constant and n the degree of non-Newtonian flow. If the flow velocity into the die is also dependent on the velocity of flow, Benbow, Oxley, and Bridgwater and Benbow and Bridgwater propose replacement of the yield value δ_y, by two empirical parameters, the initial die entry yield stress σ_{y0} and the die-entry yield-stress velocity factor α. Substituting in Eq. (18) gives Eq. (21):

$$P_0 = 2(\sigma_{y0} + \alpha \mathrm{V})\ln\ (D_0/D) \quad ...(21)$$

The fully corrected Bagley equation now becomes Eq. (22):

$$P_T = 2(\sigma_{y0} + \alpha V)\ln(D_0/D) + 2(L/R)(\tau_0 + \beta V) \quad ...(22)$$

The value of σ_{y0} can be obtained as the intercept from the derived zero-velocity graph of P_{0v0} as a function of L/R. The value of α, for a given system, is obtained from Eq. (23):

$$\alpha = (P_{0w} - P_{v0})/(2\ \ln(D_0/D)V) \quad ...(23)$$

where P_{0w} and P_{0v0} are obtained as described previously.

If the systems are treated as polymer melts instead of as paste (i.e., homogenous systems with no fluid migration during extrusion), further characterization of the wet masses can be achieved. The flow of melts through a capillary rheometer can be considered to show flow streamlines converging and then accelerating, which according to Cogswell, is extensional flow. He separated the flow field into shear and tensile deformation and then described their calculation from the following equations:

$$TS = \frac{3}{8}(n+1)P_0 \qquad ...(24)$$

where TS is the tensile stress (i.e., stretching), n is the power law index, and P_0 is the die-entrance press drop,

$$ESR = \frac{4\pi y}{3(n+1)} = \frac{\gamma}{2}\tan\theta \qquad ...(25)$$

where ESR is the tensile stretch rate, τ is the shear stress at the die wall, γ is the shear strain rate, and θ is the half angle of natural convergence.

$$EV\frac{TS}{ESR} \qquad ...(26)$$

where EV is the apparent extensional (elongational) viscosity.

Such an approach depends on the flow fitting the power law model [i.e., Eq. (10)], and that flow is not dominated by wall slip.

In addition to elongation flow, material can also exhibit elastic behavior. Two parameters that have been proposed to quantify this property are: (1) recoverable shear RS and (2) compliance C. These can be derived from:

$$RS = \frac{P}{4\pi} \qquad ...(27)$$

and

$$C = \frac{P}{4\tau^2} \qquad ...(28)$$

Chohan has used these to study the flow of branched polyethylene melt, and while what is exactly implied by these terms at high stretch rates is not clear, they are undoubtedly related to the elastic behavior of the material. The higher the values of each, the greater will be the elastic nature of the material.

Measurement of Rheological Properties

The application of the theoretical treatment depends on the ability to measure the extrusion force and rate. Most commercial extruders do not allow for these types of measurement. Normal rheological equipment, such as cup-and-bob or cone-and-plate, do not have a suitable geometry or instrumentation to handle materials of the consistency normally used. A ram extruder is a suitable experimental design.

The ram extruder, designed by Benbow and Ovenston, operates on a prefilled system and is used for experimental and small-scale extrusion. It consists of a stainless steel barrel (2.54 cm internal diameter, approximately 20 cm in length), which acts as the material reservoir. The base is constructed to enable interchangeable dies, with central capillaries of varying dimensions to be bolted on. A rubber ring is inserted between the barrel and die to ensure a water-tight connection. The piston, or ram, is a stainless steel rod that fits loosely into the barrel. A fluon ring positioned at its lower end provides a low-friction seal to prevent material escaping above the point where the piston moves down the barrel. The extrusion is a non-continuous operation; first the material (50–100 g) is packed into the barrel and partially consolidated to a plug by inserting the piston. It is possible to add temperature control to the barrel to extrude materials that are thermosensitive. The barrel-and-die assembly is mounted on a rigid metal C-piece, and a load is applied to the piston sufficient to extrude the material through the die. The ram extruder can be used in

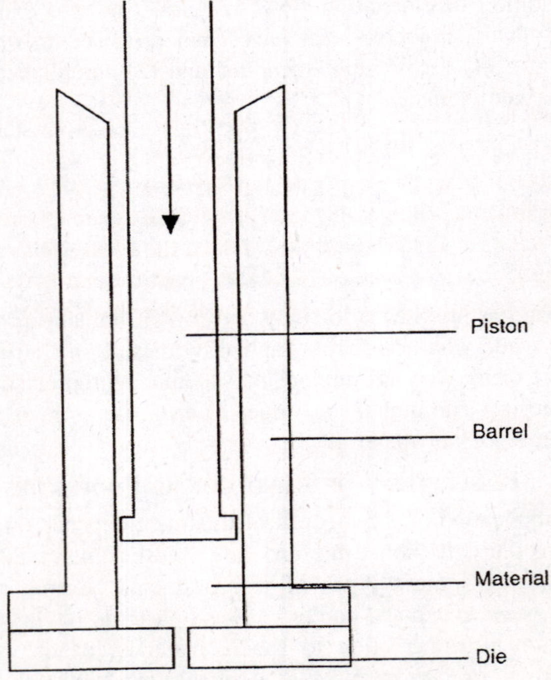

Fig. 14.1. Diagram of a ram extruder.

conjunction with an instrumented press. The piston is attached to the cross-head that may be driven down at various constant rates, and its displacement monitored by an attached displacement transducer. Output from this and the load cell is fed into an *x-y* chart recorder or computer. This arrangement enables the force acting on the material during extrusion to be recorded as a function of the displacement of the piston, and a force–displacement profile is produced.

In the compression stage, the piston descends into the barrel and consolidates the material into a plug prior to flow. This results in a large change in displacement accompanied by a small change in load. Eventually, the material is compressed to its minimum volume and maximum density. At this point, the pressure builds up while the material density is maintained. This is shown in the profile by a large increase in load accompanied by a minimal change in displacement. At the end of the compression stage, the pressure applied to the mass increases until it is high enough for the material to yield and commence flow. This is followed by a period of steady-state flow in which the force required to maintain the extrusion remains constant as the displacement increases. Forced flow occurs when steady-state flow can no longer be maintained. It leads to a gradual rise in extrusion force with displacement. This occurs often toward the end of the extrusion and is caused by the close proximity of the ram tip to the die face.

The force–displacement profile is altered by varying one of the extrusion parameters, such as the die diameter, *L*/*R*, or extrusion rate. For a given mixture, the relationship between the steady-state extrusion force, the die *L*/*R* at constant die diameter, and the extrusion rate can be represented graphically. This is known as the Bagley plot. Used in

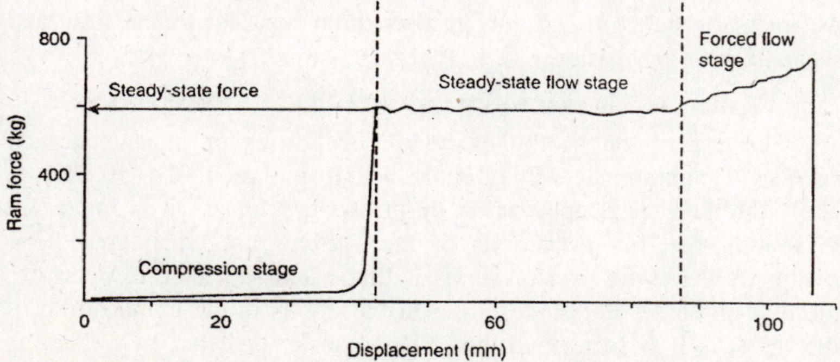

Fig. 14.2. Force-displacement profile for microcrystalline cellulose-lactose-water mixture.

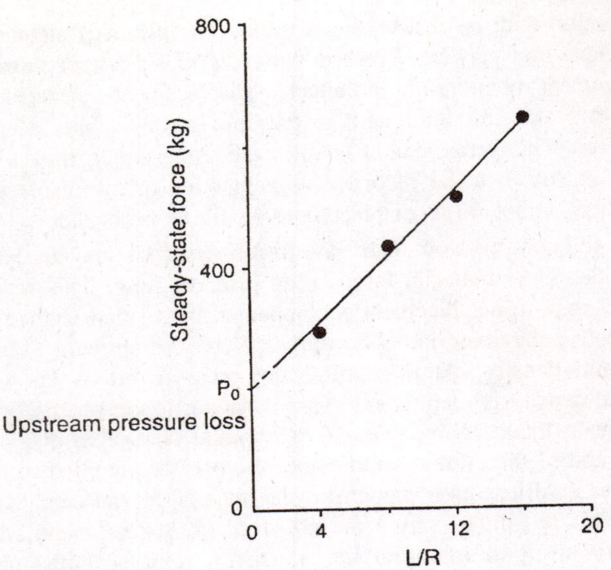

Fig. 14.3. Steady-state extrusion force as a function of the length-to-radius ratio of the die for microcrystalline cellulose-lactose-water (5:5:6) at constant die diameter (1.5 mm) and extrusion rate (20 cm/min).

this way, the extruder operates on a principle similar to that of a capillary rheometer, and expressions derived from capillary rheometry may be used to characterize the properties of the wet powder mass. After conversion of the steady-state force values to pressure values, the slope of the relationship between the pressure and L/R is numerically equivalent to twice the value of the mean die-wall shear stress. Plotting these values against the corresponding apparent die-wall shear rates results in a flow curve that is unique for a particular wet-mass formulation. The materials exhibit non-Newtonian flow and shear-thinning properties.

Practical Characterization of Extrusion Systems

The expression of the extrusion properties of pharmaceutical systems by numerical values could aid formulation. To be able to apply the theoretical approaches described previously, it is important to ensure that the restrictions of the systems are considered. One major problem with paste systems is that when subjected to pressure, there is phase separation resulting in variations in the composition of the mass as it is being extruded. This can be detected by collecting the extrudate and measuring its water content. Alternatively, magnetic

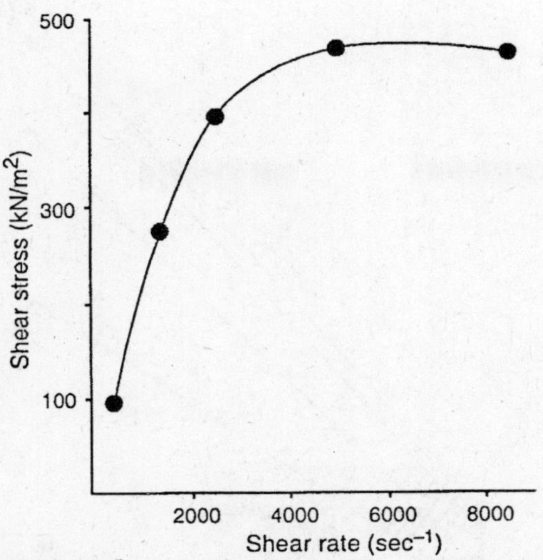

Fig. 14.4. Typical stress-shear rate flow curve for an extrusion mixture containing 50% microcrystalline cellulose extruded through a 1.5 mm diameter die.

resonance imaging has been used to quantify the water distribution within the barrel and within the extrudate.

The extent to which die-wall slip is involved can be assessed by using dies of different lengths and diameters. An important characteristic that can be observed in the extrudate is its quality in terms of surface structure. Harrison, Newton, and Rowe, have shown how this can vary from a smooth, regular surface via a rough, "*shark-skinned*" extrudate. There is obvious need to prevent this phenomenon if extrudate of the correct quality is to be produced. The occurrence of the surface defects is associated with both the composition of the material and the operating conditions (e.g., die length and diameter and the rate of extrusion). Raines, Newton, and Rowe, were able to relate the quality of the surface of the extrudate to the value of the yield stress at zero velocity $\delta_{y0,}$ in that those systems with a high value were smooth and regular while those with low values were shark-skinned.

Of the systems studied, most pastes show non-Newtonian behavior. This has important consequences for extruder design and operating conditions as material that are shear-rate dependent require careful handling. To date, most reported rheological investigations indicate that paste systems are shear thinning (i.e., their viscosity decreases with an increase in shear rate). Their extent of property can be quantified by obtaining the value of n', the slope of the log shear-rate/

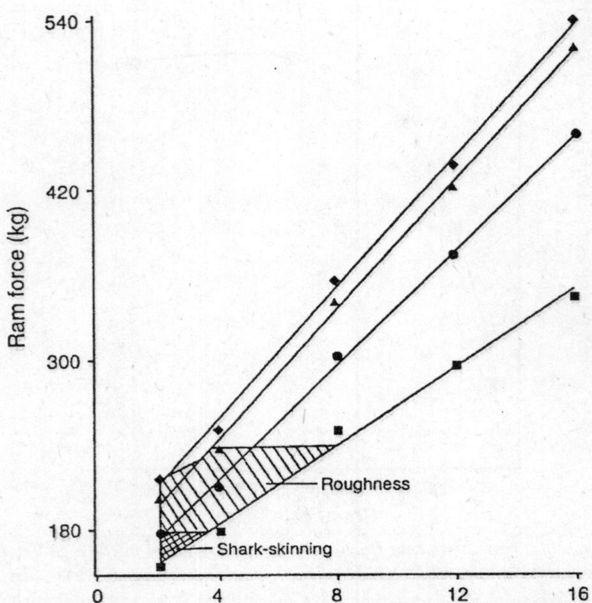

Fig. 14.5. A graph of ram force as a function of length-to-radius ratio, depicting conditions under which surface defects occur when extruding microcrystalline cellulose-lactose-water (5:5:6).

log shear-stress graph. There is also some evidence that paste systems show plug flow (i.e., the central core of the extrudate moves at a constant velocity), while there is a thin layer of moisture at the die wall where shear takes place.

Formulation

Extrusion mixtures are formulated to produce a cohesive plastic mass that remains homogeneous during extrusion. The mass must possess inherent fluidity, permitting flow during the process and self-lubricating properties as it passes through the die. The resultant extrudate must remain non-adhesive to itself and retain a degree of rigidity so that the shape imposed by the die is retained. Precise formulation requirements depend upon subsequent processing. Extrudate that is simply to be cut to short lengths to form cylindrical granules that are dried in a fluid-bed drier can be less rigid than extrudate intended for complex processing such as spheronization, where the extrudate undergoes a series of subtle shape changes.

The requirements for spheronization of the cylindrical extrudate are as follows:

1. The extrudate must possess sufficient mechanical strength when wet, yet it must be brittle enough to be broken down to short lengths in the spheronizer, but not to be so friable that it disintegrates completely. To achieve a narrow size distribution of spheres, the extrudate is ideally reduced to cylindrical rods of uniform length equal to approximately one and a half times their diameter.
2. The extrudate must be sufficiently plastic to enable the cylindrical rods to be rolled into spheres by the action of the friction plate in the spheronizer.
3. The extrudate must be non-adhesive to itself in order that each spherical granule remains discrete throughout the process.

A typical extrusion mixture might contain the following ingredients:

Drug	50–90%
Extrusion aid	
Microcrystalline cellulose, bentonite	5–50%
Binder	
Polyvinylpyrrolidone (PVP)	
Sodium carboxymethylcellulose (SCMC)	
Hydroxypropyl methylcellulose (HPMC)	
Fluid	
Water or solvent	

Extrusion offers the advantage of incorporating a relatively high proportion of active ingredient, up to 90%, in the final product. However, the physicochemical properties of the drug determine to a large extent the maximum quantity that can be included in a particular formulation. An extrusion aid is essential; microcrystalline cellulose is commonly used. The function of microcrystalline cellulose is two-fold: it controls the movement of water through the wet powder mass during extrusion, and modifies the rheological properties of the other ingredients in the mixture, conferring a degree of plasticity which allows it to be readily extruded. This interaction with the liquid phase is both a physical and chemical phenomenon. The microscopic structure of microcrystalline cellulose is a random aggregation of filamentous microcrystals that create a high internal porosity and a large surface area, approximately 130–270 m^2/g. This provides highly absorbent and moisture-retaining characteristics that are often unaffected by the extrusion process. This could be the essential quality that makes microcrystalline cellulose a unique material for extrusion. Bentonite and kaolin also have been used. Inclusion of 5–10% can significantly improve the extrusion properties of mixtures containing high proportions of drug. Recent work has shown that it is possible to reduce the quantity of microcrystalline cellulose by adding glyceryl monostearate.

Additional ingredients may or may not be necessary. A binder increases plasticity and reduces extrudate friability, particularly when the content of microcrystalline cellulose is low. Natural or synthetic polymers, such as gelatin, PVP, or SCMC, may be incorporated into the mixture as a solid during dry mixing or in solution in the liquid phase. Commercial preparations of microcrystalline cellulose that are already combined with polymers are available. Examples include Avicel RC and Avicel CL grades of microcrystalline cellulose combined with SCMC. Variations in the type of micro-crystalline cellulose significantly change the rheological properties of the mixture, and therefore, the extrusion characteristics. The differences between shear stress–shear rate flow curves of mixtures of microcrystalline cellulose–lactose–water (5:5:6) containing different particle sizes of microcrystalline cellulose and different quantities of SCMC are distinct but different. Inclusion of a polymer in the wet mass produces marked rheological differences. This has implications in the choice of formulations, since the extrudates formed from these various microcrystalline cellulose mixtures behave differently during subsequent processing, such as cutting, spheronization, and drying.

The mixture of dry ingredients is blended with water or a solvent such as ethanol to form a dense cohesive mass suitable for extrusion. The liquid content of the wet powder mass and its distribution are highly critical and should be controlled so that they produce an extrudate that posses the ideal characteristics. In general, these wet mixes have a much higher moisture content, typically 20–30 wt%, than is required for conventional (tablet) granulations, the aim being to produce as dense a material as possible for passing through the extruder. Fluffy and incompletely wetted masses feed poorly and cause problems by creating excessive pressure and friction within the equipment. On spheronizing, they tend to produce large quantities of fines, and the "dry" extrudate is insufficiently plastic, forming dumbbell-shaped or ovoid pellets which never round off into spheres. On the other hand, if the mixture is too wet, it produces an extrudate that adheres to the spheronizer plate and to itself. This product tends to aggregate uncontrollably or at best produce spheres of wide-size distribution as the material is transferred from pellet to pellet via the plate motion.

The possible processability of different drugs by this approach has not yet been fully established. It is not possible to relate the pKa, and freezing point depression, or to relate the ability to produce uniform pellets from a spheronization grade of microcrystalline cellulose. However, a relationship between the water solubility and the water level required by a formulation for equal parts of a series of model drugs and micro-crystalline cellulose has been established.

INDUSTRIAL APPLICATIONS

Plastics

Extrusion technology is extensively applied in the plastics and rubber industries where it is one of the most important fabrication processes. Examples of products made from extruded polymers include pipes, hoses, insulated wires and cables, plastic and rubber sheeting, and polystyrene tiles. The most common extruder employed is the single-screw type with either cold or hot feed, which requires the polymer to be heated prior to processing. The extruder consists of a rotating screw inside a stationary cylindrical barrel. The barrel is often manufactured in sections that are bolted or clamped together. Usually, the inner surface of the barrel is grooved to reduce slippage and increase pumping capability. An end-plate die, connected to the end of the barrel, determines the configuration of the extruded product.

The extruder is conventionally divided into three sections: feed zone, transition zone, and metering zone. Resin granules are fed from a hopper directly into the feed section, which has deeper flights or flights of greater pitch. This geometry enables the feed material to fall easily into the screw for conveying along the barrel. The pellets are transported as a solid plug to the transition zone where they are mixed, compressed, melted, and plasticized. Compression is developed by decreasing the thread pitch but maintaining a constant flight depth or by decreasing flight depth while maintaining a constant thread pitch. Both methods result in increased pressure as the material moves along the barrel. Most of the heat required to melt the material is supplied by the heat generated by friction as the resin granules are sheared between the rotating screw and the wall of the barrel. Additional heat may be supplied by electric heaters mounted on the barrel. The melt moves by circulation in a helical path by means of transverse flow, drag flow, pressure flow, and leakage: the latter two mechanisms reverse the flow of material along the barrel. The material reaches the metering zone in the form of a homogeneous plastic melt suitable for extrusion. For an extrudate of uniform thickness, flow must be

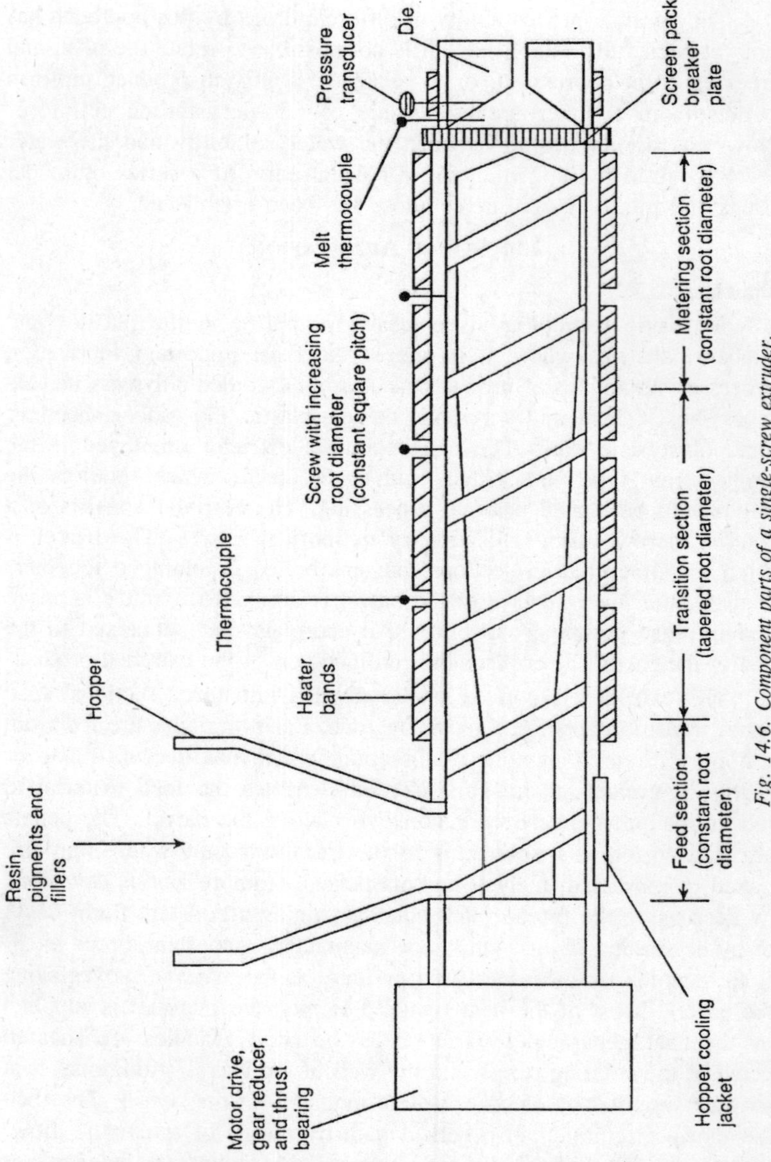

Fig. 14.6. Component parts of a single-screw extruder.

consistent and without stagnant zones right up to the die entrance. The function of the metering zone is to reduce pulsating flow and ensure a uniform delivery rate through the die cavity. Some applications require a strainer plate fitted between the extruder and die plate to remove solid impurities or lumps of incompletely melted resin.

Polymers with a wide range of viscoelastic and melt viscosities cannot be processed with a single screw. Most commercial extruders are, therefore, modular in design, providing a choice of screws or interchangeable sections that alter the configuration of the feed, transition, and metering zones. This makes it possible to modify the process to meet particular requirements, for example, from a standard to a high shear or high output extrusion. Modified screw designs allow the extruder to perform a mixing role in addition to extrusion, so that the material can be colored and blended. The various screw and die designs available and practical considerations of thermoplastic extrusion are reviewed by Whelan and Dunning. Extrusion processing requires close monitoring of the various parameters that affect polymer extrusion: viscosity, variation of viscosity with shear rate and temperature, elasticity, extensional flow, and slippage of the material over hot metal surfaces. Equations used to describe flow are included in the section on the "Theory and Characterization of Extrusion" presented earlier. Recent advances in the design and operation of extruders allow in-process monitoring and control of parameters, such as the temperature in the extruder, head, and die; pressures in extruder and die; wall thickness and other dimensions; "haul-off" speed and extrusion speed; and power consumption.

The process described above is known as profile or line extrusion in which the shape of the extrudate is determined by the die. The extruded profile proceeds horizontally to the cutoff equipment, which controls its length. It is then cooled to a solid state, usually by spraying with or immersion in water, and passed through a haul-off unit. Finally, it is cut to the required length or coiled. The downstream auxiliaries (e.g., such as haul-off equipment for handling the extrudate stream, collection machinery for winding or coiling continuous lengths of tubing or profiles, cropping and cooling equipment, and systems for monitoring the diameter and wall thicknesses of pipes on-line) are as important as the extruder itself. Tubes and pipes and other solid cross-sections are mainly produced by profile extrusion. Profiles may be further processed, for example, as in film extrusion, blow molding, or injection molding.

Film extrusion

The polymer melt is extruded through a long slit die onto highly polished cooled rolls that form and wind the finished sheet. This is known as cast film. Plastic packaging film is also formed by blow extrusion, where tubular film is produced by extruding the melt, usually vertically, through an annular-shaped slit die. The extruded tube is inflated by air to form a large cylinder. The bubble is cooled externally by an airstream directed onto its surface and is collapsed on passing between a pair of rollers before being wound up. Film made by the casting process generally has better optical properties than blown film, but is less strong mechanically. Cast films usually require edge trimming at additional cost.

Blow molding

The plastic is heated to a melted or viscous state and a section of molten polymer tubing (parison) is extruded usually downward from the die head into an open mold. The mold is closed around the parison, sealing it at one end. Compressed air is blown into the open end of the tube, expanding the viscous plastic to the walls of the cavity, thus forming the desired shape of the container. The material cools in the cavity and solidifies. The mold is opened and the molding is removed. This technique is used for the manufacture of bottles, toys, and large containers.

Injection molding

The molten plastic is extruded into a cavity mold at high pressure. The material cools in the cavity and solidifies. The mold is then opened and the article is removed. Very intricate configurations can be obtained by this technique (e.g., to provide intricate and strong components for the electronic, telecommunications, and clock-making industries).

Plastics that are commonly processed by extrusion include acrylics (polymethacrylates, polyacrylates) and copolymers of acrylonitrile; cellulosics (cellulose acetate, propionate, and acetate butyrate); polyethylene (low and high density); polypropylene; polystyrene; vinyl plastics; polycarbonates; and nylons. Additives that may be included to modify or enhance properties include lubricants and antislip agents to assist processing during extrusion; plasticizers to achieve softness and flexibility; stabilizers and antioxidants to retard or prevent degradation; and dyes and pigments.

Food

In principle, any food that can be formed into a paste can be processed by an extruder. Food extrusion has been utilized since the 1930s for pasta production. Modern equipment and processing techniques allow the manufacture of complex products in a variety of shapes and sizes. Raw materials such as cereals, oil seed, and protein, along with carbohydrates and water mixtures, can be converted into products such as meat substitutes, pet foods, and snack meals. A widely used and versatile technique combines cooking and extrusion in a so-called extrusion cooker. It has the potential to manufacture a range of novelty or specialty products, such as breakfast cereals (expanded and shaped cereals), shaped and filled snacks, protein-fortified and precooked pasta products, and precooked meat pieces for convenience foods. The process is highly economical, and provides mixing, high temperature–short duration cooking, texturizing, and shaping of the food in one step. The equipment closely resembles the screw extruders used in the processing of thermoplastics. The screw is designed to create varying zones along the barrel, allowing the food substance to be processed in stages. The solid and liquid starting materials are fed from a hopper to the feed zone of the extruder and conveyed to the transition zone. Here the materials may be compressed, mixed, sheared, and heated to form a viscous plastic dough. In the metering zone, the plastic mass is subjected to further heating and shearing before being pumped into the die to form the shaped product. The pressure drop on leaving the die causes superheated water to flash off the molten material. If the dough contains starch, gelatinization will result in an expanded porous product with a crunchy texture. Finally, the product may be cut, shaped by passing through rollers, dried, and packed.

The viscosity of the dough may vary more than an order of magnitude during the extrusion cooking process as a result of changes in shear rate, temperature, moisture content, and induced physicochemical changes such as protein denaturation, polysaccharide gel formation, and reorientation of molecules. For this reason, success in food extrusion requires accurate monitoring and control of feed rate, screw speed, temperature, and moisture to produce and control desired product characteristics. Knowledge of the viscous rheology of the food mixture in the metering section immediately prior to extrusion is of particular importance. However, this is not easily predicted since, unlike the case of homogeneous or simple mixtures of polymers where the major change is melting, food doughs are of such complexity that the exact chemical composition and structure cannot readily be determined. Efforts have been made to develop semiempirical models derived from plastics extrusion to describe the apparent viscosity of cooking doughs, which may be useful in evaluating food formulations.

Remsen and Clark used an Instron capillary viscometer and amylograph to describe the relationship between the viscosity of a typical soy flour dough and the applied shear rate, temperature, and time–temperature history. Fletcher et al. investigated the viscous dough rheology of maize mixtures as a function of the extrusion variables (pressure, shear, and temperature). They used an instrumented single-screw extruder fitted with slit dies, and related the results to the product properties. The advantage with this method is that the food material receives a deformation history corresponding to the extrusion cooking process, which is otherwise difficult to replicate in a laboratory rheometer.

Animal Feed Production

In the animal feed industry, extrusion is applied as a means of producing pelletized feeds, commonly in the form of short cylindrical rods of 4–8 mm in diameter. Pellets are a convenient means of precisely controlling the animal's diet. They offer several advantages. The quantity of feed the animal receives is better controlled by pellets than a loose-mix feed, and a complex diet of controlled composition is easily produced. The pellet feed can contain as many as 30 single ingredients mixed in the correct proportions. The animal is obliged to chew pellet feed with improved palatability and therefore, digestion. During extrusion, the feed mixture is compressed, resulting in a densified product that requires less storage space.

Pellets are prepared from a mixture of raw materials of varying chemical composition (starch, oil, fiber, and moisture) and physical characteristics (particle size, bulk density, and moisture-retention properties). The composition of a typical poultry feed is often complex. The raw material properties determine the quality of pellet formation. Equipment performance and pellet quality can be improved by a small amount of extrusion or pelleting aid. Additives commonly used in the feed industry include molasses with binding properties when activated by steam; fatty acid lubricants to reduce product–metal friction when extruding or pressing through long dies; lignosulfates (organic materials derived from lignin in trees that improve pellet quality and throughput rates); and mineral binders, such as ball clay and bentonite, or cellulose binders, such as sodium carboxymethylcellulose. It should be noted that in small quantities cellulose binders can improve the pelleting process and reduce pellet friability.

The pelleting process consists of blending and conditioning the feed mixture immediately prior to pressing, pressing itself, cutting of the pellets, and cooling. The complexity of the feed mixture, composed of a number of ingredients of different particle sizes and densities in varying proportions, requires thorough blending to ensure homogeneity. The product is conditioned by adding moisture, typically up to 15%, and heating to a controlled temperature in order to gelatinize the starch or convert it to simple sugars. This reaction causes the starch to act as a binder and converts the meal into a physical state suitable for pressing. The most efficient means of conditioning and heating is by steam. Optimal conditioning parameters, moisture content of the material, temperature, and duration of heating depend on the composition of the mixture. For example, high starch–low fiber meals require temperatures of 80–85°C, whereas feeds that contain heat-sensitive ingredients, such as milk and sugar, have a temperature limit of 55°C.

Extrusion Pressing

According to Sebestyen, pellet mills may be classified into disk-die presses or ring-die pellet mills. In the former, the die consists of a circular plate resting in the horizontal plane into which holes are drilled in a regular pattern. A set of rollers move around the upper surface of the disk, sweeping the meal in their path through the holes and compressing it to form pellets or cubes. Rotating adjustable knives located beneath the disk cut the extrudate to an appropriate length. In another design, the plate revolves while the rollers and knives remain fixed. Ring-type pellet mills have a radially arranged die resting in the horizontal plane with rollers rotating and revolving along the inner surface. The rolls are offset from the die face, leaving a slight clearance that allows buildup of a thin product layer, optimizing throughput efficiency. The peripheral velocity of the rollers depends upon the die diameter; that is, higher speeds are required for smaller-diameter holes and lower speeds for larger-diameter holes.

Cooling

On leaving the extruder, the warm pellets are pliable and prone to abrasion and deformation. Therefore, a final processing stage is required to harden the pellets. Cooling equipment placed directly beneath the mill employs ambient or chilled air to reduce the temperature and remove excess moisture from the final product.

Pharmaceutical Industry

Extrusion processes are applied within the pharmaceutical industry to produce a variety of dosage forms such as suppositories, implants, and granulations.

The large-scale manufacture of suppositories and pessaries uses either the fusion method where the drug is dispersed in a molten base and the mixture poured into molds to solidify, or the cold compression method.[43,44]In the latter process, the medicament and cold-grated base, usually theobroma oil or witepsol base, are intimately mixed and placed in a cylinder. The mass is extruded by means of a piston through small holes that connect with the mold. The cavities are filled by pressure with the mass which is prevented from escaping by movable end plates. The plates are removed and the suppositories ejected by further extrusion. The extrusion equipment is chilled to prevent melting of the components due to the heat generated by the friction of compression.

The most important application of extrusion in the pharmaceutical industry is in the preparation of granules or pellets of uniform size, shape, and density that contain one or more drugs. The process involves a preliminary stage in which dry powders, drug, and excipients are mixed by conventional blenders, followed by addition of a liquid phase and further mixing to ensure homogeneous distribution. The wet powder mass is extruded through cylindrical dies or perforated screens with circular holes, typically 0.5–2.0 mm in diameter, to form cylindrical extrudates. These may be further processed, for example, by cutting and drying to yield granules, or by spheronization to yield spherical granules followed by drying. The spheroids are usually coated with a polymer to control the rate of drug release and filled into hard gelatin capsules to yield a multiple-unit dosage form.

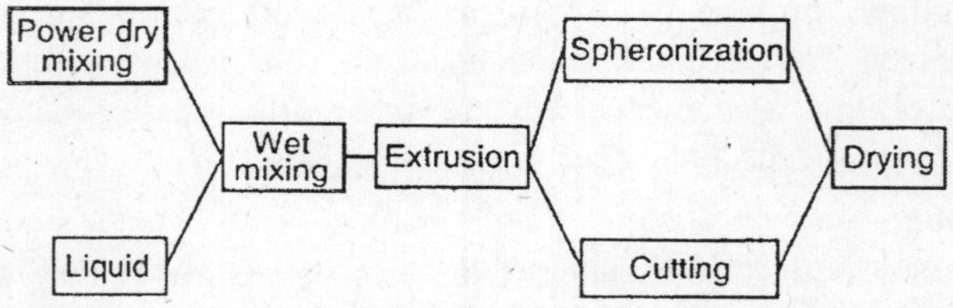

Fig. 14.7. Schematic of extrusion processing in the pharmaceutical industry.

Extruders Used for Pharmaceuticals

Commercial extruders may be classified according to the die design and the feed mechanism that transports the material to the die region.

Screen extruders

Screen extruders utilize a screw-feed mechanism consisting of single or twin helical screws rotating in a barrel to convey the damp mass from a feed hopper to the die zone. The die consists of a thin

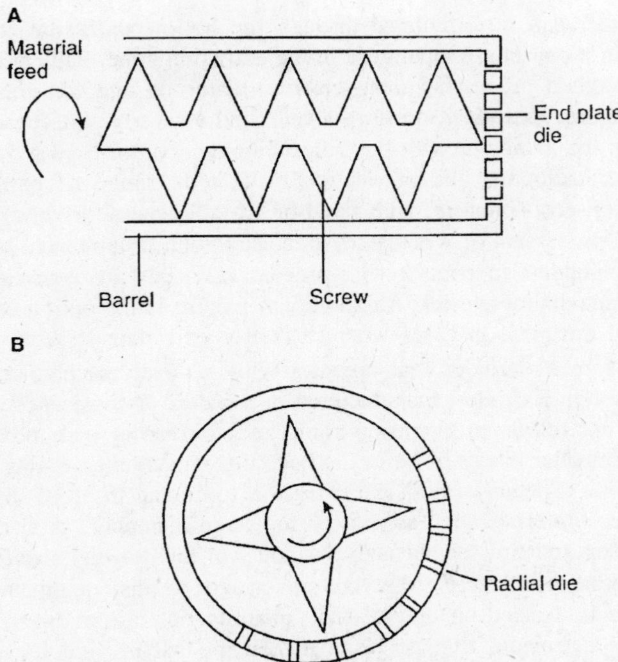

Fig. 14.8. Screw extruder with (A) end-plate die and (B) radial screen die.

steel plate perforated with numerous holes, which is positioned radially or axially to the screw feed. The advantages of this arrangement are high continuous throughput rates, from 5 kg/h of wet mass for a laboratory-scale single-screw extruder, up to 800 kg/h for a larger twin-screw design. The screens are easily cleaned and interchanged; they have holes of varying diameter beginning at 0.5 mm and are available commercially. The disadvantage of this type of equipment, however, is that the screw mechanism can exert a high pressure on the material, generating excessive friction and heat as the wet mass passes between the screw and barrel. This is particularly the case with axially orientated dies. These extruders tend to have a high dead volume that contains stagnant material between the feed screws and the screen. Consideration should be given to this if the wet powder mass contains ingredients that are unstable when wetted with water. The low *L*/*R* of the die holes can also result in low compaction in the extrudate and distortion of the surface finish, known as shark-skinning. This problem can sometimes be overcome by varying the throughput rate, which will be discussed later.

Water can be circulated through the hollow extrusion rotors to maintain a constant temperature in the extrusion zone. This is a useful facility when processing heat-sensitive materials and for controlling temperature, extrudate moisture levels, and viscosity. Interchangeable screens are available with die holes ranging from 0.5 to 1.5 mm in diameter, allowing the production of a wide range of extrudates. Explosion-proof motors, with fixed or variable speed drive, are fitted for safe processing of wet masses granulated with inflammable solvents. All components in contact with process materials are constructed of high-grade stainless steel. An additional feature is the option of fitting an axial die plate in cases where a denser extrudate is required.

A screen extruder that operates with a novel mechanism is the Nica System Extruder, manufactured in Sweden. It consists of a radial screen encircling an extrusion rotor and a rotating disk fitted with angled impeller blades or baffles. Above this is a counterrotating central feed blade. Speeds of both the extrusion rotor and the feed blade are variable. Material, such as gravity fed from a hopper, is swept into the blades and pressed through the holes in the screen, according to the manufacturer, there are several advantages to this equipment. First, pressure is exerted on only a small quantity of mass and only at the point of extrusion; that is, just between the baffles and the screen. Second, temperature increase is minimal, and a moisture gradient between the wet mass and extrudate is avoided. Because of this, cooling facilities are not necessary. The dead volume, located in front of each baffle, is limited and may be as low as 15 g per baffle. A small extruder with output up to 4 kg/min is available for development and small-scale production. For larger production, an extruder with output of up to 12 kg/min is available.

Rotary-cylinder extruder

The working principle of this machine is based on two counterrotating cylinders. The granulating cylinder is perforated and acts as the die. The diameter and the *L/R* of the holes can be varied. The holes are spaced further apart and are drilled rather than of punched sheet construction as in the screen-type extruders. The other cylinder is solid and acts as a pressure cylinder. Material is gravity fed from a hopper to the die region between the cylinders and adheres to the knurled surface of the solid cylinder, building up a thin layer that is pressed through the die cylinder. Although the extrusion is a continuous process, actual material flow though each hole is intermittent due to the rotation of the die. Pressure is built up in the perforations,

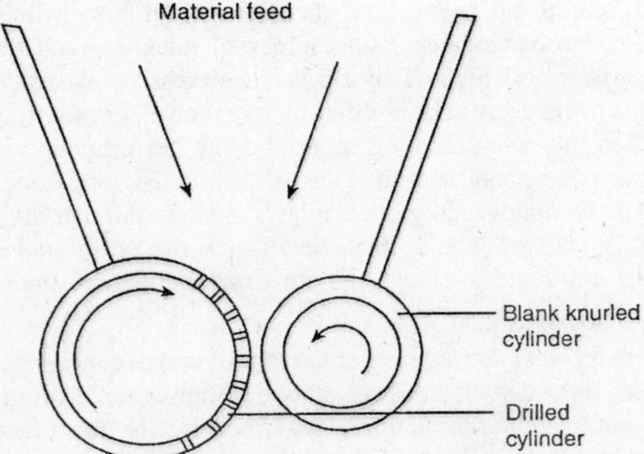

Fig. 14.9. The rotary-cylinder-type extruder.

which compacts the wet mass and forces the extrudate to the interior of the cylinder. This pressure is dependent upon the diameter and the length of the perforations. Hence, with die holes of high *L*/*R*, this system can achieve good densification of the wet powder mass. This is important in giving the most granules mechanical strength and stability for further processing. Another advantage is the lack of a dead zone, which is limited to the thin layer of material adhering to the pressure cylinder. Since the cylinders only apply pressure to a small quantity of material in the feed zone, there is little tendency for creating moisture gradients. However, cleaning of the granulating cylinder can be troublesome. The material remaining in the die holes can be difficult to dislodge, particularly when the L/R is high. Furthermore, the granulating cylinders are expensive because of the high costs of drilling stainless steel.

With cylinder extruder, the feed stock material can be metered to the working area. On the smallest machines this is accomplished byarotary-table feed hopper. On larger machines, the feed rate is controlled by screw feeders sited through the feed hopper. The throughput rate depends on the diameter and *L*/*R* of the die holes, as well as on the feed rate. Laboratory-scale extruders with a throughput range of 30 to 50 kg/h use granulation cylinders 70 mm in diameter. Production-scale equipment with a larger granulating cylinder (186 mm diameter) can achieve anoutput of 100–105 kg/h. Interchangeable cylinders with die holes of 1.0–5.0 mm are available. A multiple-unit assembly consisting of three parallel extrusion heads can achieve even higher throughput rates of up to 3000 kg/h.

When scaling up production, the die cylinder dimensions cannot be increased proportionately, since the wall thickness and therefore, die *L/R* become too high. This results in excessive extrusion forces imparted on the material. This is overcome by using special counterbored die cylinders with reduced depth perforations to provide optimal extrusion conditions at scale-up. The temperature increase in the extrudate is minimized by circulating cool water through the pressure cylinders. A scraper blade attachment inside the perforated cylinder cuts off the extrudate to shorter lengths, making it more manageable.

Rotary-gear extruder

The rotary-gear extruder operates on a similar concept to that of the cylinder extruder. It consists of two hollow counterrotating gear cylinders with counterbored dies, described by the manufacturer as nozzles, that are drilled into the cylinders between the teeth. The material, gravity fed from a hopper, is drawn in by the toothed cylinders and pushed through the nozzles into the center of the cylinders, where scrapers cut off the extrudate. The product is compacted as it passes through the nozzles, and thereby forms a dense extrudate. The density depends on the nozzle *L/D* (the ratio of nozzle length to nozzle diameter). Higher throughput rates can be attained with this type of extruder, since output is achieved through both rotating-gear wheels.

Interchangeable gear cylinders are available with variable nozzle *L/D* ratios by counterboring from the inside of the rollers to reduce the die length or by using replaceable nozzle inserts to increase the die length. The diameter of the holes can be varied from 1.0 to 10.0 mm to produce a range of pellet sizes. The through-put rate can be controlled by varying the cylinders' rotation speed and the corresponding feed rate. Throughput capacity ranges from 20 kg/h for the small-scale laboratory extruders to approximately 1000 kg/h for production equipment. For large equipment or materials with poor flow characteristics, agitators can be installed in the hopper to prevent bridging; special hoppers with conical feed screws fitted with additional wipers are used for highly viscous products. The machine can be furnished with cooling equipment, circulating water through the compaction gears for processing materials that need temperature control.

Ram extruder

Industrial ram extruders are commonly used in the plastics and rubber industries for the preparation of warm strip feed for large cold-fed screw extruders and for forming strips or slugs for feeding injection molding and compression molding machines. They are used in the extrusion of specialized substances that require critical in-process control or that are not readily amenable to processing by screw extruders. Examples include the extrusion of waxlike substances such as coloring crayons, dental waxes, and rocket propellant, and in the extrusion of moist powders and claylike materials, such as blackboard chalks. Ram extrusion allows control of parameters, such as temperature, size, and weight of extrudate. It consists of a chrome-plated barrel positioned in a thermostatically controlled storage tunnel. A range of dies can be fitted to the extruder head. Material is loaded into the barrel by manual or mechanical means and vacuum is applied to eliminate air from the system.

The material is extruded by means of an hydraulically powered ram, with the hydraulic (oil) fluid being passed through a special valve system to sense changes in the plasticity of the material and compensate ram pressure to achieve an even extrusion through the hole. A multi speed cutter mounted on a fly wheel severs the extrudate at the die face. The volume of the extrudate is a function of the cut speed and the set ram speed, and can be controlled to a high degree of accuracy within ±1%. A continuous operation is possible with the help of a twin barrel and ram arrangement in which material is fed to each barrel in turn by a screw system. Various sizes of extruder are available, from 4.5-L barrel capacity up to 80 L, offering production rates up to a maximum of 800 kg/h.

Choice of extruder

The selection of the extruder design is based on the principal requirements of the extrudate and the nature of further processing. For the production of uniform granules to be dried in a fluid-bed drier, a low-compaction system, such as that provided by the various types of screen extruders may be suitable. Cylinder or gear-type extruders may be more appropriate when aiming for a densified extrudate, such as that required for spheronization. Ram-extrusion systems, which allow precision control of extrudate density, size, and shape, are ideal for the extrusion and forming of pharmaceutical polymers of the type used for sub-dermal implants.

Consideration should be given to the availability of small-scale equipment, which is vital for development work prior to scale-up on pilot- or production-scale machines. Equipment choice is not necessarily based on maximum throughput rate, since the subsequent processing stages (e.g., cutting, spheronization, and drying) are batch processes and are therefore, a rate- limiting factor in production. Since extrusion is a continuous process, it allows adequate production rates for most purposes with any of the above mentioned extruder types.

The equipment must comply with the code of Good Manufacturing Practice (GMP) standards within the pharmaceutical industry. Machines should be constructed of durable material with smooth surfaces to discourage adhesion of extraneous material and facilitate cleaning. Materials used for equipment construction must not affect the product. They should be corrosion resistant and able to withstand cleaning disinfectants. All surfaces and parts in contact with the product must therefore be of high-grade stainless steel and designed to exclude contamination of process material or product by lubricant during manufacture. Controls should be accessible by means of recessed contact buttons.

15

Transgenic Crops

It has now been 16 years since the first mammalian protein was expressed in plants. The concept of *plant recombinant protein* (PRP) production developed from that early experiment. The first reports of PRPs were published in the late 1980s when antibodies were produced in tobacco and human serum albumin was expressed in tobacco and potato. That work has now evolved into a burgeoning industry, which aims to produce a myriad of protein-based industrial and biopharmaceutical products in crop plants, aquatic plants and algae. Crop plants are particularly suitable for the production of recombinant proteins because they offer what amounts to infinite scalability and low up-front production costs. Unlike fermentation systems, growing plants are not constrained by physical facilities. Recombinant proteins that had only been available in microgram quantities can now be produced in plants by the kilogram. However, if the advantages of scale are to be realized, then crops expressing genes for a wide range of recombinant proteins must leave the greenhouse for evaluation in confined field trials.

Regulation of Field-testing

Field-testing is a critical component of the commercialization and scale-up process for agricultural biotechnology. It provides an opportunity to examine the interaction between the transgene and the production conditions, to examine the environmental impact of the plants, to develop production projections and GMP processes, and to create material for pilot scale manufacturing, pre-clinical and early clinical experimentations. Field-testing played a major role in the introduction of the first generation of input traits from agricultural biotechnology, such as herbicide and insect resistance. More than 10,000 transgenic field trials have been conducted since 1989 and they provided insight into efficacy and agronomic performance for a wide range of transgenic crops. Field-testing has been similarly important in the development of crops engineered for recombinant protein production, although there were few trials until the mid 1990s. As a result of those PRP field trials, two products (β-glucuronidase and avidin), both of which are laboratory reagents, are now commercially available. There is no large-scale commercial production of any PRP crop thus far, but as the many potential products move through the development process, large-scale production will follow.

In most countries, a permit is required to grow a genetically modified crop in the field. Typically, genetically modified means that a recombinant DNA technique has been used in the creation of that crop, but the exact definition does vary by country. In Canada the permitting body for field trials is the Canadian Food Inspection Agency (CFIA). In the United States it is the United States Department of Agriculture's Animal and Plant Health Inspection Service (USDA-APHIS), and similar organizations exist in Europe, Asia and South America. Since 1990 there have been 246 field trial applications worldwide involving the production of industrial or biopharmaceutical proteins, with 76% of those trials conducted in the United States, 20% in Canada, 3% in Europe and 1% in Argentina. The number of applications to field test PRP crops peaked in North America between 1995 and 1996 and again between 1999 and 2001. Following those peaks, the number of field trial applications in North America again began to decline, resulting in a cycling pattern of trial numbers. In Canada, the latest decline was attributed to a "pause" in field testing during 2001, while the CFIA developed new regulatory protocols for PRP crops.

The Canadian moratorium has now been lifted and field test numbers have rebounded. In the United States, the current decline may reflect normal fluctuations in the product development cycle, which have been seen in previous years, or they may result from newly recognized liability issues. That liability is an issue became apparent during what has become known as the "*Prodigene Affair*". In the fall of 2002, a small number of volunteer corn plants appeared in a soybean crop in the year following a field trial involving transgenic corn containing a biopharmaceutical. The trial was conducted by a Texas based crop biotechnology company called *Prodigene*. The soybeans were harvested and may have contained traces of the dry tissue from the transgenic corn plants that grew in the field. As a part of their field trial permit requirements, Prodigene was required to have removed any volunteer corn plants, but failed to do so and received a substantial fine as a consequence. Although risk to the public was probably minimal, the incident did bring risks associated with PRP production to the forefront for public debate.

The regulation of crops producing pharmaceutical or industrial proteins was governed by generic directives in both the United States and Canada until 2003. The philosophy underlying the North America system is the regulation of the product and not the process, whereas in Europe it is the opposite. In either case the information required for field trials by the various regulatory agencies is similar. For field testing PRP crops, proponents submitted applications that contained information relating to the unmodified plant, the modified plant, the construct used in the transformation, reproductive isolation, the nature of the site and its potential environmental impact, trial protocol, and post-harvest land use and monitoring.

Each application was considered on a case-by-case basis and was assessed for environmental safety. Field trials for PRP crops typically required only environment safety assessment, which included such things as potential for weediness or invasiveness, gene flow to wild relatives, and non-target impacts or impacts on biodiversity. If the material produced in the trial was for use in early clinical experiments, then efficacy and safety would be evaluated by other regulatory agencies such as Heath Canada or the Food and Drug Administration (FDA) in the United States. Otherwise these agencies are not involved in the field-testing of PRP crops. In-depth examinations of the principles used to regulate genetically modified crops around the world and the key issues associated with the assessment of ecological risk are provided in two recent reviews.

Containment is essential for the field production of PRP crops, as was first identified by Menassa and co-workers who made a strong case for the use of male-sterile tobacco in order to keep PRP products out of the human food chain. Recombinant proteins produced for medical or industrial use can have untoward biological activity in humans and other animals, and the risks of either acute or chronic exposure are often unknown. Therefore, it is critical to keep crops producing such materials out of the food and feed chain, and the environment. Commandeur *et al.* have detailed the risk factors specifically associated with the production of PRPs and made a distinction between the ecological/health risk and those associated with product safety. They divided the risk associated with PRP crops into two categories: the first related to the spread of the transgene beyond the platform crop leading ultimately to human exposure and the second was product safety, which is the risk that the products of molecular farming *per se* were harmful to humans or other animals. They suggested numerous means to reduce the risk of transgene escape including the use of self-pollinated, non-food crops with minimum exposure to wild relatives along with various means to minimize the amount of genetic material introduced into the crop.

The issues surrounding PRP crop production were discussed in depth at a recent Pew Initiative Workshop focused on the benefits and risks of plant recombinant pharmaceutical production. The major benefit identified was the capability of plants to produce low-cost recombinant proteins. As a result of favorable costs, plant-based production systems could be used to meet the growing demand for new protein based medicines, particularly those that are too cost-prohibitive when produced in conventional mammalian cell culture systems. Risks associated with the escape of the PRP crop were recognized and concerns were expressed about the ability of industry to contain these crops and prevent their entry into the food chain. USDA regulators pointed to new, more stringent, field-testing regulations for PRP crops and indicated that PRP crops will be the subject of regulatory oversight throughout the research and commercial production cycle, all of which would serve to ensure containment.

From a field testing and field production standpoint, the risk profile for transgenic crops producing biopharmaceuticals and even industrial enzymes is quite different from that of the first generation transgenic crops now in unconfined release. Risk is the product of probability and consequences. Under a standard set of production conditions the probability of an adverse event, like pollen dispersal for example, is the same for PRP transgenics as it is for transgenics carrying first generation traits. However, the consequences of that event may be far worse, given that the recombinant protein may be biologically active in humans, and therefore the risk is higher. Recognizing that the risk level of PRP crop field trials using standard production methods is higher than that associated with conventional transgenic crops, both the US and Canadian regulatory authorities deviated from the use of generic regulations for field tests involving PRP crops, and in 2003 issued amended regulations. In the United States, USDA-APHIS strengthened their permit conditions by: requiring more inspection visits, increasing the amount of information required about harvest and storage procedures, implementing mandatory training programs for staff involved with the permit, increasing buffer zones from 800 to 1600 m for corn-based production, restricting the production of food and feed crops at the test site in the post-harvest growing season, increasing the size of the fallow zone from 7.5 to 15 m and requiring the use of dedicated equipment for harvest.

Miller was critical of the changes and felt that they were unnecessarily restrictive. He argued that the new regulations were the result of the Prodigene affair and that the resulting real risk to consumers was "*extremely low*". Others, like the Union of Concerned Scientists, were calling for more stringent measures such as disallowing food crops as production platforms, indoor production, mandatory male sterility and designated grow zones.

In Canada the interim regulatory amendments governing field trials with "plants with novel traits" that produce industrial or pharmaceutical proteins were more conservative than those issued in the United States. For example, the CFIA is recommending that major food or feed species, and crops pollinated by honeybees, are not to be used for PRP production. The use of non-food, fiber, small acreage or new crops is encouraged. They asked that proponents consider the potential for escape of the crop or the transgene into the environment when choosing a crop platform. Plants used to produce PRPs should be "*amenable to confinement*". Isolation distances were increased, and the cultivation of food and feed crops following a PRP crop was discouraged. New hazard and exposure data for human and livestock health assessment may also be required from PRP-containing traditional food or feed crops prior to the approval of field trials. Exposure risk concerns the potential for PRPs to be present in human food or animal feed, and where exposure can occur, what mechanisms are used to limit biological activity. Hazards included direct toxicity and allergenicity in humans or animals as well as hazards presented by the co-product streams that result from processing. These latter requirements could place a major burden on proponents to prove their materials are safe prior to even confined field trials.

Design of Field Trials

There are three considerations in the design of field trials for PRP crops. The first is eliminating harm to human health, the second is mitigation of harm to the environment and the third is experimental design. Human exposure could result from accidental release of the material, which can be eliminated through careful material handling procedures and record keeping. In Canada for example, inspectors must witness the destruction of residual plant material and records of the dispensation of retained material must be kept.

Other issues include the exposure of workers involved in the trial and they can be resolved through the use of protective gear or by harvest equipment design. Ensuring that the environment is not harmed requires a clear understanding of the biology of the host crop species. Documents outlining taxonomy, genetics, reproductive biology, potential for inter-specific hybridization, weediness potential and field production systems are available from the regulatory authorities in Canada and the United States. These documents help to determine areas of potential risk during the production cycle and proponents then develop means of limiting that risk. For example, although tobacco is not an out-crossing species, it is a prolific seed producer and trial protocols often include the removal of flowers or harvest prior to flowering to limit the potential for seeds to escape into the environment. Physical isolation is another method of limiting escape.

In Canada and the United States, isolation distances for various crops used in the production of recombinant proteins are prescribed. Isolation can also be achieved by conducting trials outside of normal production zones for a given crop, although the feasibility of such a procedure is questionable, since the agricultural environment providing the isolation may not be suitable because it is isolated from the normal production zone. In addition to physical distance, male-sterility has been used in tobacco to re-enforce containment during the field-testing of tobacco plants producing proteins in their leaves. Seed-based production systems such as corn have also used sterility-based containment for field-testing. Many molecular containment methods have been developed but none have as yet been deployed in PRP crop field trials.

From a research perspective, field trials, like other types of experimentation, are designed around a hypothesis. For example, the yield of recombinant protein from a selected transgenic line is theoretically 300 kg ha^{-1}. Therefore, one may hypothesize that the yield will be the same in agricultural production and a procedure is designed to validate the hypothesis. There are four phases in this process, i.e. the selection of: test materials, traits to measure, measurement procedures, and data analysis methods. In our example, the test materials would be any number of transgenic lines selected for high recombinant protein yields. The traits to measure could be the yield of target tissue or the concentration of recombinant protein, which together would provide the yield of recombinant protein per unit area.

The remaining two phases represent the experimental design stage and allow experimental error to be estimated and controlled, and the results of the experiment to be interpreted. In field trials, experimental error is estimated through the use of replication and randomization. Multiple plots containing the same treatments, and assignments of plots at random to locations within the trial, allow the real differences between the treatments to be separated from experimental error. In this example, the trial should be repeated in space and time to provide a more accurate representation of recombinant protein yield in a normal range of agricultural conditions. The use of a process called *blocking* to account for environmental gradients in the field, along with optimum plot management and methods of data analysis help to control experimental error and enhance the researcher's ability to see treatment differences.

Results of Field Trials

The few studies that have been published to date tend to include field trials as a part of large pilot studies aiming to demonstrate the feasibility of producing a particular protein in a given system. All of the trials used some reproductive measure like male-sterility or flower removal to control the spread of pollen. The results of multi- location experiments with corn expressing avidin were among the first to be published. The trials were conducted throughout the United States and made use of off-season nurseries to speed up the trial process. The authors concluded that avidin accumulation in corn was stable across the multiple environments, although generations and locations appeared to be confounded.

Gastric lipase was produced in a field test in France, but the focus of the experiments was extraction efficiency and characterization, and the results of field performance were not presented. Tobacco was used to produce human interleukin-10 in a single location field trial conducted in Canada. The authors examined the performance of five male sterile transgenic lines in an effort to estimate IL-10 production levels and to determine if male sterility was affecting crop vigor. Male sterility had no affect on biomass yield and IL-10 was produced at 0.5–1.0 g ha^{-1}. Lee and co-workers used field trials of transgenic white clover expressing a vaccine antigen-GFP fusion protein in order to produce material to test for storage stability. The antigen present in the dried clover was found to be stable for as long as six months.

There has been a large number of field trials of transgenic crops that accumulate recombinant proteins, but the results of only a few have been reported so far. Of particular interest for the future are reports of detailed trial protocols that address confinement concerns; multi-location, multi-year performance and stability data; environmental impact and non-target impact studies; and GLP production. The impact of production conditions on extraction efficiency and product quality are also significant issues that should be addressed through field-testing.

16

PARTICLE ENGINEERING

Particle engineering is a term coined to encompass means of producing particles having a defined morphology, particle size distribution, and composition. In its broadest sense, wet granulation can be considered a means of particle engineering, as it is a way of transforming a diverse set of particle sizes into uniform conglomerates. In a stricter sense, particle engineering is associated with particle size reduction techniques, such as media milling and homogenization, and micro- or nanoparticle formation techniques, such as spray-drying, supercritical fluid technologies, and precipitation. Recent additions to the pharmaceutical particle engineering arsenal include spray-freezing into cryogenic liquids, template emulsion, and self- assembly. This chapter will concentrate on those technologies involving drug particle size reduction or formation in the unit micrometer to nanometer range, with references from the recent literature, but it is by no means inclusive.

POTENTIAL ADVANTAGES OF PARTICLE ENGINEERED DRUGS

Improved Oral Absorption

Between 1995 and 2003, only 9% of new drug applications to the FDA described pharmaceuticals with high solubility and high permeability. As the number of poorly soluble drugs continues to rise, particle engineering technologies are being considered for reducing size and/or crystallinity of drugs in order to improve oral absorption. If the particle size is reduced, an improvement in oral absorption is expected for Biopharmaceutics Classification Class II compounds, that is, those which have poor solubility and high permeability. Such compounds have dissolution-rate limited absorption as opposed to permeability-rate limited absorption. The improvement in bioavailability is attributed to the decrease in particle size leading to an increase in the surface area, A, which leads to an increase in the rate of dissolution, as per the Noyes–Whitney equation:

$$\frac{dM}{dt} = kA(C_s - C) \quad \text{...(1)}$$

where M = mass dissolved, t = time, k = intrinsic dissolution rate constant = D/h, where D is the diffusion coefficient, h is the thickness of the diffusion layer, C_s is the saturation solubility of the drug in the diffusion layer, and C is the concentration of the drug in the bulk solution.

Improving the rate of dissolution for highly permeable drugs may result in an improvement in absorption, provided there are no other hurdles, such as metabolism or paraglycoprotein efflux.

In this simple model, drug particles in the solid state are in equilibrium with drug in solution. When the concentration of drug in solution is less than the saturation solubility, the equilibrium is shifted to the right and favors dissolution of the solid. The rate of dissolution is affected by the particle size.

Assuming that the drug must be in solution to be absorbed, if the equilibrium is driven to the right by reducing particle size, absorption should improve, providing the absorption rate constant, k_a, is not rate-limiting. If absorption is the rate-limiting step, a supersaturated state may result, which may drive the equilibrium to the left, resulting in "*crashing out*," that is, precipitation or recrystallization in uncontrolled particle size. This may result in dissolution slow down and poor or erratic absorption.

The rate of dissolution of drug in the amorphous state may exceed that in the crystalline state and in the nanocrystalline state, resulting in an instantaneous solubility which may exceed the equilibrium solubility. In such a case, if k_a is not rate-limiting, the fraction absorbed may exceed that obtained by simply reducing particle size.

Content Uniformity

In cases whereby a drug is very potent, and drug loading in a final dosage form is very low, the drug must be of sufficiently small a particle size in order to ensure content uniformity of the dosage form. If the particle size is too large, no amount of blending can ensure content uniformity. Particle formation technologies such as SEDS [solution enhanced dispersion by supercritical (SC) fluids], spray-drying,

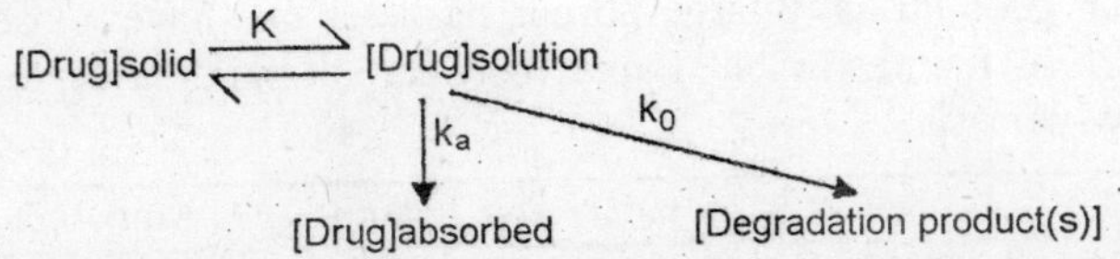

and wet milling or homogenization are capable of producing particles of a narrow particle size distribution with a mean diameter under 5 μm. Conventional dry milling techniques cannot always reproducibly achieve such a size and distribution. Additionally, dry milling can impart a high-energy amorphous surface, which can lead to high static charge and particle cohesion. These surface attributes can lead to processing issues during wet granulation, including lack of content uniformity, enhanced drug degradation, and potential for polymorphism.

Pulmonary Delivery

Drugs for pulmonary administration can be administered via dry powder inhaler or by inhalation of an aerosolized or vaporized formulation. In order to deliver a drug to the deep lung, particles must not be large or heavy enough to remain in the back of the throat or in the upper part of the lung, where absorption cannot occur. On the other hand, they cannot be so small or light as to be exhaled immediately. In addition, the formulation must be sterile.

Quantitatively, the appropriateness of particles for pulmonary delivery is assessed by the aerodynamic diameter, a term developed by aerosol physicists which takes into account size, shape, and density, and is typically determined by an inertial sampling device such as a cascade impactor. The aerodynamic diameter, D_{aer} (μm), for particles greater than 500 nm can be approximated by:

$$D_{aer} = D_S \sqrt{\rho} \qquad \ldots(2)$$

where D_S is the Stokes, or geometric, diameter (μm), and ρ is the particle density (g/cm^3). Non-spherical particles, such as fibers and agglomerates, which appear to have different physical sizes and shapes, may have the same aerodynamic diameter:

	ρ (g/cm^3)	D_S (μm)	D_{aer} (μm)
Solid sphere	2	1.4	2
Hollow sphere	0.5	2.8	2
Elongated irregular solid	2.4	1.3	2

Likewise, particles which appear to be visually of similar size and shape may have different aerodynamic diameter if their densities vary, and geometrically large porous particles can have aerodynamic diameters in the respirable range (between 1 and 5 μm), owing to their low density:

	ρ (g/cm^3)	D_s (μm)	D_{aer} (μm)
Low-density irregular sphere	1	2	2.0
Medium-density irregular sphere	2	2	2.8
High-density irregular sphere	3	2	3.5
Large, porous particle	0.02	30	4.2

By varying process parameters and feedstock composition, spray-drying and supercritical fluid processing can be used to prepare particles of various morphologies, densities, and size in a very narrow particle size distribution. It is not surprising, therefore, to find many examples in the literature where these two techniques have been used to process pure drug, including proteins and peptides, or drug with an excipient in order to produce particles for pulmonary delivery. Other particle reduction/formation techniques, such as microfluidization, spray-freeze drying, and other precipitation techniques have also been used for this purpose, in addition to the classical approach of air-jet milling.

Interestingly, large porous particles have been found to be more appropriate than smaller, dense particles for pulmonary delivery. This is because particles with D in the respirable size range and a density close to 1 are subject to rapid, macrophage uptake and clearance in about 30 min, whereas geometrically large particles are avoided by this clearance mechanism. Controlled release of insulin over several days was demonstrated by this technique using a poly-lactide-co-glycolide (PLGA) copolymer matrix; however, drug loading was very low. Concerns regarding toxicity of PLGA oligomers limit the usefulness of this matrix; however, variants of this approach can be expected in the future.

It should be kept in mind that the list of approved excipients for delivery to the lung is very limited. Excipients added to dry powder inhalers, such as lactose, are meant to be of large particle size and to be deposited in the back of the throat, whereas the smaller-sized drug is that which reaches the lung. For those technologies resulting in an inhalable matrix of drug and an excipient, additional toxicology tests may be needed to qualify the excipient.

Chemical Stabilization of Proteins and Small Molecules

A drug which is prone to hydrolysis in aqueous solution may be stabilized by formulating it as an aqueous suspension or nanosuspension.

An aqueous nanosuspension is inherently more stable to hydrolysis than a solution, because only the portion in solution is available for degradation, much like ordinary suspensions. Thus, for a drug which has a solubility of 0.01 mg/ml, and is formulated as a 10mg/ml suspension, only 0.01/10 or 1/1000 of the dose is available for hydrolysis at any time t. As the drug in solution degrades, the equilibrium is shifted to the right, toward dissolution of the suspension. As long as the solid remains, the drug concentration in solution remains essentially constant and equal to the saturation solubility of the drug in the vehicle. This leads to an apparent zero-order, as opposed to first order, rate of degradation, and is typical for suspensions, that is, a plot of drug concentration vs. time decreases linearly as opposed to by an exponential decay:

Zero order:

$$C_t = C_0 - k_0 t \qquad ...(3)$$

First order:

$$\ln C = C_0 - k_1 f \qquad ...(4)$$

where C_t = concentration of undecomposed drug at time t, C_0 = initial drug concentration, k_0 is the zero-order rate constant, and k_1 is the first-order rate constant.

If a drug is subject to degradation in the solid state as opposed to hydrolysis, the greater surface area of particles in a nanosuspension may render it less stable than an ordinary suspension.

Proteins can be stabilized by coprocessing them with sugars, sugar alcohols, or polymers by techniques such as spray-drying, supercritical fluid processing, or lyophilization. Typically, the goal is to increase the T_g, or glass transition temperature, of the protein such that it remains in the glassy state at room temperature, as dictated by the Gordon–Taylor equation:

$$T_{g,\max} = \frac{w_1 T_{g1} + K_{w2} T_{g2}}{w_1 + K_{w2}} \qquad ...(5)$$

where T_{g1}, T_{g2} and $T_{g,mix}$ are the glass transition temperatures (°K) of components 1, 2, and the mixture, and w_1 and w_2 are the weight fractions of each component. The constant K is approximated as

$$K \approx \frac{\rho_1 T_{g1}}{\rho_2 T_{g2}} \qquad ...(6)$$

where ρ_1 and ρ_2 are the densities of each component.

In this immobilized state, proteins are more stable than in the rubbery state above the T_g. Interestingly, enzymatic activity can be retained by processing techniques such as spray-drying, even though it involves heat, because spray-drying is an evaporative cooling process.

Intravenous Delivery

Intravenous delivery of poorly water-soluble drugs is especially challenging, as it is most preferable to deliver the drug as a solution in order to avoid the formation of pulmonary emboli or induce a hypotensive effect. Additionally, it must be demonstrated that aggregation does not occur upon contact with blood components.

Delivery of particulates intravenously is possible as long as the particles are less than 5–7 μm, which is the size of the smallest pulmonary capillary. Particle engineering technologies such as wet bead milling and homogenization, as well as some of the spray-based technologies, are capable of providing particles small enough for intravenous administration. This was demonstrated in an in vivo tumor model, where it was found that 100 nm paclitaxel nanoparticles manufactured by wet bead milling and injected i.v. had a remission rate at day 70 of 28%, compared to 12% for 300 nm nanoparticles injected i.v., and 0% for the commercial product injected i.v. Successful intravenous administration was also demonstrated in a 2-week clinical trial.

Thus far, most particulate administration has been restricted to infrequently dosed pharmaceuticals, such as phase contrast agents. Repeat intravenous administration of particulates is a young science which is expected to benefit from emerging particle engineering technologies.

Targeted Delivery

Uptake of intact nanoparticulates from the *gastrointestinal tract* (GIT) after oral administration, for example, by endocytosis by the M cells of Peyer's patches, is an area of controversy. Most research shows that the efficiency of this process is extremely low (much less than 1%), the extent is poorly defined and subject to great variability. Rather, it is the enhanced rate of dissolution of nanoparticulates which results in improved exposure.

Nanosized drug particles administered parenterally, on the other hand, are naturally cleared by the *mononuclear phagocyte system* (MPS) by a process known as *opsonization*. When particulates such as cell fragments, macromolecular proteins, and microorganisms enter the bloodstream, they are quickly cleared by phagocytic cells of the MPS, such as Kupffer cells in the liver. The opsonization process involves adsorption of plasma proteins, such as complement fragment C3b, onto the particles, followed by phagocytosis by the macrophages in the liver (60–90%), the spleen (1–5%) and, to a minor extent, in the lung. The exploitation of the opsonization process to target organs such as the liver and the spleen by using colloidal drug carrier systems has been an area of intense investigation for many years. This approach has also been investigated as a means to treat infections such as tuberculosis, which are caused by parasites residing in the MPS.

Alternately, nanoparticles can be designed to avoid clearance by the MPS so as to increase blood circulation time. Prolonged circulation of drug nanoparticles in the blood may increase the drug's probability of being deposited near a tumor via the so-called "*leaky*" vasculature, that is, by extravasation (leaking), by diffusion or convection, through the fenestrated endothelia of capillaries near tumors.

Distribution of nanoparticulates after parenteral administration appears to be governed by similar mechanisms as are liposomes, mainly size and surface characteristics such as charge and hydrophilicity. For example, liposomes, which are also cleared by the MPS, are more likely to reach and accumulate in tumors if the size is between 90 and 200 nm.

Another means of avoiding recognition and clearance by plasma proteins is through the use of steric, a.k.a. stealth, particle surface modifiers or stabilizers such as *polyethylene glycol* (PEG). The hydrophilic nature of the PEG chains render them protein-repelling. For example, PEGylated phospholipids have been used to make liposomes stealth.

In vivo, PEGylated polymeric poly(lactic acid) PLA nanoparticles have been shown to have reduced uptake by the MPS, and a slow, constant extended release of drug in the circulation, with a resultant increase in AUC.

In addition to PEGylation, polyvinyl pyrolidone (PVP) nanoparticles modified with hydrophilic polaxamine have been reported to adsorb less protein than the unmodified particles, conventional liposomes, and stealth liposomes. Polystyrene microspheres coated with lecithin and particles coated with Pluronic were also observed to adsorb less protein and demonstrate a prolonged blood circulation time without adversely affecting the safety profile of the drug.

Intraperitoneal or intramuscular delivery of particulates is known to result in lymphatic uptake, and may be a means of targeting lymph nodes local to the injection site in a more efficient manner than intravenous administration.

It may be possible to use drug nanoparticles to target other non-MPS targets, such as the bone marrow or the brain without concomitant entry of toxins into the central nervous system. For example, the peptide dalargin was successfully targeted to the brain using polyisobutlycyanoacrylate (PBCA) nanoparticles which were surface modified with Tween 80. The Tween 80 coated PBCA particles showed preferential adsorption of Apo-E, which is implicated as the mechanism leading to brain targeting.

For all these approaches, the dilution effect afforded by the circulatory system upon intravenous administration, and by the GIT upon oral administration, may result in an ability to modulate the pharmacokinetics depending on the solubility of the drug and the concentration of the administered dose. If the dose is low relative to the solubility, the nanoparticles may become solubilized and behave as a solution. However, if the dose is high, the dosage form may remain in the nanoparticulate form for a longer period of time, which may lead to a sustained release effect.

Ocular and Topical Delivery

One of the main challenges of ocular delivery in man is the constant production of tears, which, coupled with blinking, effectively minimizes residence time of applied pharmaceuticals. Additionally, the epithelial layers of the cornea, and to a lesser extent the sclera and conjunctiva, provide barriers to exogenous compounds. Although the pharmacopeia may allow up to 75 μm sized particles in ophthalmic suspensions, in order to minimize potential irritation to the eye and reduce elimination of the instilled drug through tearing, the particle size of the drug should be less than 10 μm and be spherical in shape. Particle engineering techniques can produce such particles.

Pegylated PLA nanospheres, solid lipid nanoparticles, and chitosan nanoparticles have been reported to be well-tolerated ocular drug carriers in rabbit models with an improved rate of efficacy. Confocal microscopy suggests the chitosan nanoparticles penetrate into the corneal and conjunctival epithelia, enhancing residence time without being cytotoxic. Residence time has also been enhanced in rabbits by coating poly-epsiloncaprolactone biodegradable *nanospheres* (NS) with bioadhesives such as hyaluronic acid, which is an endogenous component of the eye. Cationic surfactants used in the production of the NS include benzalkonium chloride, which is approved for use in ophthalmics, and stearylamine.

Solid dispersion particles made by spray-drying disulfiram 1:1 or 1:2 (w/w) with PVP and instilled in the rabbit eye were found to have improved concentrations in the aqueous humor compared to 1:1 solid dispersions made by the evaporation method. In both cases, drug particles were passed through a 75 μm sieve. Particle size of the spray-dried particles had a D_{50} of 3.3 $\pm$ 0.04 μm, while those prepared by evaporation had a D_{50} of 34.3 $\pm$ 18 μm.

Solid lipid nanoparticles (SLN) are considered to be the next generation of topical delivery following liposomes. The excipients are well tolerated, and the SLN have good adhesive and film-forming properties, which may provide an enhanced occlusive effect. SLN have improved stability toward hydrolysis, and show potential for modulated drug release. SLN have been found to scatter UV light when applied topically, and hence provide a sun-protective effect, similar to the mechanism of particulate titanium dioxide. Molecular sunscreens or UV-blockers incorporated into SLNs show a synergistic sun-blocking effect while at the same time reducing side effects such as skin penetration and consequent skin irritation.

Particle Engineering Technologies

Grinding Techniques

Mechanical grinding of drugs in the dry or wet state has been used to reduce particle size, and, if desired, crystallinity. Conventional dry milling techniques, such as micronization by air jet milling, which can produce particles with mean sizes in the unit micrometer or larger, will not be discussed in detail here.

Dry ball milling (high-energy mechanochemical activation)

This decades-old technology is mostly used to reduce crystallinity as well as particle size in order to improve bioavailability. In general, drug is ball milled for hours in the dry state or in the presence of solvent vapors. Particle size reduction occurs, but typically not to the nano size range. Drug can be dry-milled alone or in the presence of surface modifying agents which prevent the newly formed surfaces from agglomerating, stabilize the amorphous state, and/or form inclusion complexes with the drug. Stabilizers include β-1,4-glucan, β-cyclodextrin, cross-linked carboxymethylcellulose, crospovidone, silica gel, neusilin, or dextran. Typical drug–excipient ratios (by weight) are 1:1 to 1:3, or in the case of cyclodextrin, 1:9. This process may enable certain chemical reactions leading to degradation or in some cases explosion, and thus must be monitored for such, as would be any dry milling technique.

Nanosuspensions by wet bead milling

The preparation of pharmaceutical nanosuspensions by wet (bead, or media) milling reduces drug particles to a mean particle size less than 0.4 μm, more commonly 100 or 200 nm. In this process, drug is wet milled as a suspension in water, or in a medium such as safflower oil, ethanol, *t*-butanol, hexane, or glycol. The milling dispersion also contains up to several percent of one or more surface modifiers which adsorb onto the freshly formed surfaces of the drug and prevent agglomeration through steric and/or electrostatic stabilization. Typical surface modifiers include low-viscosity pluronics, PVPs, hydroxypropyl celluloses, hydroxypropyl methylcelluloses, PEGs, lecithin, dextran, aerosol OT, polyethylene glycol, Tween 80, sodium lauryl sulfate, docusate sodium, and sodium deoxycholate. Typical drug to excipient ratios range from 20:1 to 1:1.

Milling is accomplished in the presence of yttrium stabilized zirconium oxide beads or highly cross-linked plastic beads, such as polystyrene, or glass beads, ranging in size from 0.2 to 3 mm, in a media mill, such as a DynoMill or NanoMill. At the end of milling, the nanosuspension is separated from the beads prior to administration or further processing. Drug concentrations as high as 40% can be milled, and milling time can range from minutes to days depending on the energy put into the milling process. Since this is an attrition process, milling efficiency is improved by milling at high rather than low drug concentrations. Milling can be done under refrigerated conditions which minimize thermal degradation.

The drug in nanosuspensions made by wet bead milling is predominantly crystalline. Properly formulated nanosuspensions are physically stable under ambient or refrigerated conditions for months to years, and have the appearance and viscosity of milk. It is possible to convert the nanosuspension into a dry form by spray-drying, spray granulation, spray-coating onto cellulose or sugar spheres, or lyophilization. The dry form can then be further processed into capsules or tablets in which the drug redisperses to the nano size in the presence of an aqueous environment.

Residual monomer is typically less than 50 ppb, and insoluble residue generated during milling is no more than 0.005% (w/w) based on the drug concentration of the dispersion or resulting dosage form.

High-pressure homogenization

Homogenization is a technology which has been in use on a large scale since the 1950s. Nanosuspensions and SLNs can be produced by

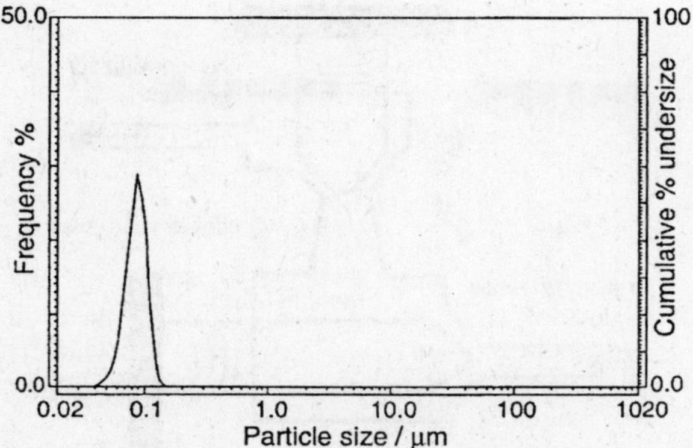

Fig. 16.1. Example of a particle size volume distribution obtained by bead milling, and determined using a Horiba LA-910, using water as the dispersant, a relative refractive index of 1.20 - 0.10i, in the standard mode.

homogenization using piston-gap homogenizers or microfluidizers. These processes generate heat, and may expose the product to increases in temperature of 10–20°C per cycle. If lipids or other low-melting excipients or drugs are involved, the melting point needs to be taken into consideration.

Homogenization is a technique applied to rupture cells, including bacteria, and thus has a sterilizing effect on the drug product. Metal contamination from the homogenization process has been found to be very low. The most dominant contaminant, iron, was detected as less than 1 ppm after 20 cycles at a maximum pressure of 1500 bar.

Nanosuspensions by homogenization

Processes involving high pressure and high shear, such as piston-gap homogenization and microfluidization, can be used to produce nanosuspensions. The major difference between homogenization and wet milling is the absence of milling media (i.e., beads), which must be removed from the final product.

Drug, preferably jet-milled, is first dispersed in a surfactant solution using a high-speed or overhead stirrer, at concentrations ranging from 1% to 15% solids content. In order to minimize plugging of the gap during the high-pressure homogenization process, the particle size of a coarse suspension can be reduced by first passing it through a low-pressure cycle at 100 or 150 bar. During the high-pressure cycles, the drug suspension is passed through a very small gap, for example, a 25 μm gap at 1500 bar. This leads to cavitation, or formation of bubbles which implode when the suspension leaves the gap and normal

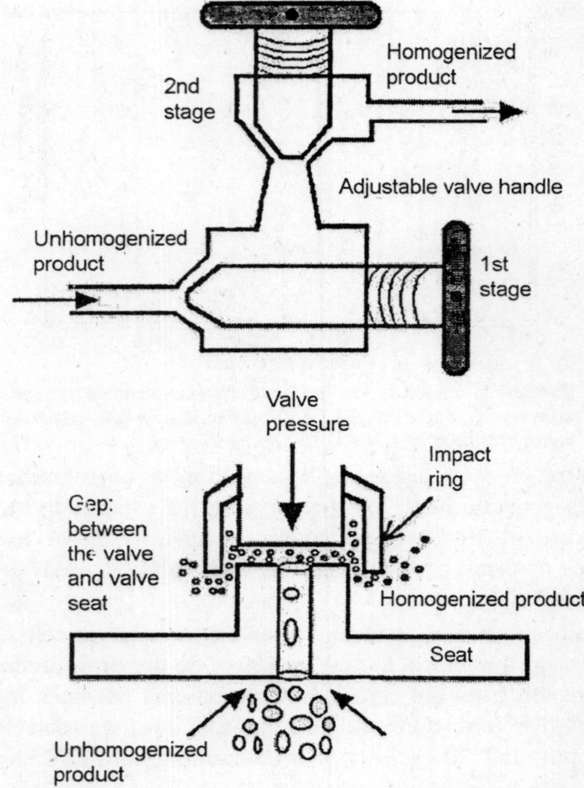

Fig. 16.2. Diagram of operation of a piston gap homogenizer.

air pressure is reached again. Cavitation coupled with high-shear forces in the gap and particle collision result in particle size reduction. Typically, 3–20 cycles are required to fully reduce the drug to nanoparticle range. High-pressure homogenization may lead to a decrease in crystallinity of the drug.

The piston gap homogenizers are made in such a way that the drug suspension is hurled against homogenizer surfaces, while microfiuidizers are arranged such that the nanosuspension stream is split and then the two streams reunited to impinge against each under high pressure (2500–25,000 psi). The latter is said to be more efficient and requires fewer passes to reach the desired end point. To prevent excess heating of the sample during homogenization, the liquid stream is circulated through a heat exchange coil to keep the product at refrigerated temperatures.

A typical formulation might include up to 2% polaxamer 188, up to 0.5% polysorbate 80, up to 0.1% sodium cholate, and up to 1% lecithin, or 3–10% phospholipids or lecithin. Typically, lipids are more easily processed by homogenization than by bead milling. Nanosuspensions can be preserved with, for example, benzyl alcohol, or methyl paraben and propyl paraben.

A modification of this technique first precipitates the drug, dissolved in an organic solvent, into an anti-solvent, which can be water, which is then homogenized; stabilizers may be added to either phase. Stabilizers/crystal growth inhibitors include PVP, a cellulosic polymer, polaxamer, or a membrane-forming lipid, such as a phospholipid. Alternately, the drug solution in organic solvent is mixed into an oil-in-water emulsion composed of, for example, cottonseed oil, PVP, methyl oleate, and water; drug is absorbed into the emulsion droplets. The precipitated stabilized particles or emulsion may be further size-reduced by ultra high-pressure homogenization. Alternately, the water-miscible solvent such as methanol, is stripped by evaporation, and standard techniques such as spray-drying or lyophilization are used to remove water and isolate dry particles.

In a variation of this theme, the stabilizer used is albumin. This technology results in an amorphous nanoparticle form of the drug coated by albumin, and having a size of about 100–200 nm. Drugs prepared by this technology have been tested by intravenous, intra-arterial, inhalational, oral, and topical administration. In the case of i.v. paclitaxel, the albumin is said to result in an increased and prolonged intracellular availability of the drug.

Solid lipid nanoparticles

Solid lipid nanoparticles were first described in 1991 as an alternative to emulsions, liposomes, and polymeric nanoparticles for intravenous administration. The latter, typically prepared from the biodegradable polyesters PLA and PLGA, were found to be difficult to scale and also cytotoxic due to oligomeric degradation products formed when the polymers were internalized by macrophages. Solid lipid nanoparticles were found to be much less cytotoxic in cell cultures: 100% mortality was observed from incubation with 0.5% PLGA nanoparticles, while viability remained at 80% in 10% SLN. Solid lipid nanoparticles of paclitaxel and camtothecin injected intravenously showed higher and prolonged plasma levels and increased uptake by the brain. Interestingly, both stealth and control SLNs of paclitaxel showed low uptake by the liver and spleen.

In order to be incorporated into an SLN, a drug must be soluble in the lipid phase. Drug loading has been reported to be between 5% and 25%. Lipid micro- or nanoparticles can be produced by spray congealing, but large-scale production is most easily and cost-effectively accomplished by homogenization.

Typical solid lipids used are glycerides and/or fatty acids, and may constitute 30% of the formulation. These are from the same family of lipids found in parenteral nutrition emulsions, such as Intralipid, which have been successfully administered intravenously for several decades. Typical excipients are Dynasan 112, composed of short chain fatty acids, Compritol, lecithin, used as an emulsifler, and surfactants such as polysorbate 80, polaxamer 188, PVP, bile salts such as sodium glycocholate, and Span 85. Water can be replaced with oils or PEG 600 to yield dispersions which can be filled into soft gelatin capsules.

The process generally involves preparation of a thermodynamically stable emulsion consisting of the drug, water, surfactant, cosurfactant, and molten lipid, which is homogenized in a piston-gap homogenizer then cooled to recrystallize the lipid, leading to solid lipid nanoparticles. Alternately, a drug/lipid melt is cooled, ground to about 50–100 μm, dispersed in a cold surfactant, and homogenized at or below room temperature. Final particle size of the SLNs range from 50 nm to 1 μm in a narrow particle size distribution.

Solid lipid nanoparticles can be converted to solid dosage forms by, for example, granulation, whereby the SLN dispersion is used as the granulating fluid, or as the wetting agent in the extrusion process, or by spray-drying or lyophilizing the SLN dispersion.

Spraying Techniques

Spray-drying (ambient temperature or above)

Spray-drying has been used in the pharmaceutical industry for over 60 years, as it offers a one-step process for converting a solution, emulsion, or suspension into dry particles which can be crystalline and in the unit micron size range. Aqueous or organic solvent feeds can be utilized, however, for the latter; the system must be made inert by the use of nitrogen in order to avoid explosion.

Much of the pivotal research into spray-drying was done by Masters in the 1950s, and his handbook remains one of the most useful references for those working in the field. The process has enjoyed enormous success, and has been scaled up for the production of milk products, coffee, vitamins, drugs such as acetominophen, and directly compressible excipients such as lactose and microcrystalline cellulose. Additionally, spray-drying has been used to convert nanoparticle dispersions into solid dosage forms.

The feedstock is atomized by a rotary wheel or nozzle to form a spray of droplets. These come into contact with a hot air or gas such as nitrogen, which rapidly dries the liquid. Because spray-drying is an evaporative cooling process, proteins and thermally labile drugs such as antibiotics can be processed with minimal loss in activity or potency.

Variables affecting the size, morphology, density, crystallinity, and polymorph of the final particles include droplet size, feedstock total solids concentration, rate of atomization, inlet temperature, product temperature, and residence time (dictated by the geometry of the spray dryer). In some cases, cospray-drying drug with a polymer such as PVP or a high molecular weight PEG can lead to stabilized amorphous particles containing drug and polymer.

Depending on the operating conditions the resultant particles can be hollow or solid, with a smooth or wrinkled surface, and of high or low density. In microspheres, drug is homogenously dispersed into a solid excipient matrix as fine particles or agglomerates, or in the molecular state. In the case of englobing (solid in solid) or microencapsulation (liquid in solid), the surface of the droplet dries first, forming a skin surrounding a vapor bubble. As drying continues, pressure in the core builds and is relieved by an eruption through the skin, yielding hollow interiors, and a skin containing the trapped oil or solid drug. For hollow particles, a film forming excipient, such as gum arabic, gelatin, and maltodextrin, is included.

Spray-drying proteins such as interferon with less than 2% stabilizers such as human serum albumin, and/or bulking agents such as carbohydrates (lactose, trehalose, raffinose, maltodextrins, mannitol), polypeptides (such as aspartame), amino acids (such as alanine and glycine), and/or buffer salts (such as sodium citrate) have been found to provide stable compositions suitable for use in dry powder inhalers. The method of production involves stabilizing an amorphous form of the drug by coprocessing with, for example, a sugar, which raises the glass transition temperature (T_g) of the drug, rendering the amorphous form of the drug in the particles more physically stable. Typically, aqueous solutions adjusted to the pH of greatest stability for the protein would be used as the feedstock, and spray-drying conditions would be relatively mild (inlet temperature less than 100°C), to avoid thermal deactivation of the protein.

Spray-freezing into cryogenic liquids

This process produces stabilized highly porous amorphous nanostructured particles with improved wetting and rate of dissolution. The crystallinity can be tuned by varying the processing solvents, stabilizers, and conditions. Drug loading has been reported to be as high as 91%.

In this process, a feedstock containing drug (for example, 2%), with or without excipient(s) such as Pluronic F 127, nucleation inhibitors such as PVP K10 or K15, polyvinyl alcohol, or hydroxypropyl-beta-cyclodextrin (HP-β-CD), is atomized below the surface of compressed liquids such as CO_2, helium, propane, ethane, or cryogenic (temperatures below –40°C) liquids such as nitrogen, argon, or hydrofluoroethers, and the droplets are immediately frozen. The solvent in the frozen particles is then sublimed by lyophilization or atmospheric freeze-drying. The latter technique fluidizes the frozen powder using cryogenic air or nitrogen. The cryogenic air or nitrogen is then heated to ambient temperature, yielding dried micronized powder. Drug–excipient ratios are typically between 1:2 and 2:1. The feedstock solution, suspension, or emulsion can be aqueous, organic (for example, ethanol, acetonitrile, or tetrahydrofuran), or a mixture.

Resultant particles have a high glass transition temperature, which aids in stabilizing the amorphous state. These particles can be compressed into tablets. In one case, tablets containing 150 mg drug were pressed from a 1:1 mixture of the processed drug powder, microcrystalline cellulose and 5% carboxymethylcellulose.

Supercritical fluid technologies

In a temperature vs. pressure phase diagram, the triple point is defined as the point at which solid, gas, and liquid coexist as three separate phases. If one follows the liquid/gas coexistence line above the triple point, a critical point is reached at a specific temperature and pressure, at which a single phase is observed. The SC fluid region exists above this point, and is a single phase. In this region, compounds such as carbon dioxide (CO_2), water, and ethanol have the viscosity and diffusivity of a gas, but the densities and solvent power of a liquid, making SC fluids ideal for extraction. Unlike pure liquids and gases, small changes in temperature or pressure can result in large changes in density, thus solvent power. This can be manipulated to affect the rate of nucleation, and hence size, density, crystallinity and polymorph, of the resultant particles. Because CO_2 becomes SC at the relatively mild conditions of 31.1°C and 73.8 bar (or 7.38 MPa) of pressure, and because CO_2 is inexpensive, relatively non-polluting, and non-flammable, it is the most frequently used supercritical fluid. Supercritical CO_2 is considered to be as non-polar as hexane, yet to be slightly acidic.

The advantage over conventional crystallization processes is that recrystallization and particle size reduction can be accomplished in a single step with minimal operator exposure and minimal use of organic solvents.

Supercritical fluids have been used for extraction for over 100 years, and industrial-scale extraction plants are in operation for the decaffeination of coffee, and extraction of hops, spices, herbs, and other natural raw materials. This is a popular technique as the extracts are of high quality and solvent-free. The extraction power of SC fluids was eventually applied to the production of solvent-free fine particles of drug of defined size distribution in the mid-1980s. This is accomplished by aerosolizing the SC fluid, a solution of drug, or both through the use of a nozzle or capillary. As with spray- drying, the size of the droplet is an important variable in determining the particle size.

Fine particle production by SC fluid technology can be broadly classified into two types, one in which the drug is soluble in the SC fluid, and one in which the drug is not. In both cases, drug is first dissolved in a suitable solvent which is soluble in SC CO_2. In the expansion step, the pressure is reduced to atmospheric during which the SC fluid becomes a gas, and extracts with it the solvent, leaving particles behind. Excipients can be included in the process.

For drugs which are soluble in SC CO_2, the SC fluid can be expanded into air, or into a solution containing polymers and other additives, such as phospholipids or Tween 80. By varying depressurization, the product can be formed as very fine particles, fibers, thin films, or as biodegradable polymeric microspheres containing drug. Typical biodegradable polymers which are soluble in the SC fluid include poly(hydroxyacid) and poly(D,L-lactic acid). Scale-up has been complicated by nozzle blockage and aggregation of particles during the expansion process.

Most drugs and proteins are not soluble in commonly used supercritical fluids, and therefore are processed instead by the SC antisolvent technique, the most popularized being the SEDS. SEDS-produced crystals can have extremely smooth surfaces, as shown by scanning electron microscopy and atomic force microscopy, and the surface may be more hydrophobic and less wettable than crystals grown under more polar conditions.

It is possible to coprocess drug with excipients to make particles in which the drug is either homogenously dispersed in an excipient matrix or else coated by it. Polymers used are those not appreciably soluble in SC CO_2. Coating is of interest for extended release, taste-masking, protection from moisture, light, or oxygen, and to improve flow properties. Typical coating agents include the Eudragits. To avoid agglomeration, the operating temperature is maintained below the T_g of the coating polymer. In the case of DNA, addition of dissolved mannitol in the buffered aqueous feed allowed recovery of at least 80% of the original supercoiled DNA.

Other Precipitation Technologies

Solvent based processes, based on precipitation techniques other than SC, have been used to prepare drug microspheres and nanoparticles for many decades. Precipitation techniques generally rely on drug being soluble in at least one solvent which must be miscible with an anti-solvent, or that drug be insoluble within a certain pH and/or temperature range. The resultant aqueous colloidal dispersions may be converted to a dry state by, for example, lyophilization or spray-drying, or centrifugation and tray or fluid bed drying.

Impinging jet is an example of classical solvent/ antisolvent technology, whereby a solvent solution of drug is impinged at high pressure with an antisolvent, resulting in rapid precipitation of small particles, typically in the micrometer size range. Additionally, an excipient can be incorporated into either the solvent or the antisolvent stream. The resultant particles can be dried by spray drying or by centrifugation and tray or fluid bed drying.

Bioerodable protein microsphere technology relies on precipitation of proteins in the presence of water- soluble polymers by changing pH, ionic strength, protein and polymer concentration, as well as temperature, and mixing speed. Stabilizing polymers include PEG, PVP, dextran, or hetastarch. No organic solvents are used, which distinguishes them from classical polymer microspheres which are prepared by homogenization of a solvent-based emulsion. The microspheres are collected by centrifugation and lyophilized. Proteins and peptides tested have included albumin, hGH, insulin, and leuprolide acetate. Particles formed are very homogeneous and of a very narrow particle size distribution in the 1–2 μm size range, suitable for inhalation, or in the

20 μm range, suitable for parenteral administration (other than intravenous). Drug loading of 20–90% protein has been reported.

Evaporative precipitation into aqueous solution (EPAS) involves dissolving the drug in a low boiling organic solvent such as methylene chloride or diethyl ether, then heating the solvent under pressure to above the boiling point. The liquid is atomized into a heated aqueous antisolvent solution containing stabilizing polymers and surfactants. Stabilizers and surfactants interact with the drug nuclei as they are being formed through hydrogen or hydrophobic bonding, and prevent further growth or agglomeration. The stabilized aqueous nanosuspension can be lyophilized or spray-dried, or spray-granulated in a fluid bed processor onto substrate, for example, an 80:20 blend of lactose and microcrystalline cellulose. Drug loadings as high as 90% have been reported. Stabilizers include PVP 40T, PVP K-15, alone, or combined with SLS, pluronic, or sodium deoxycholate to aid in wetting, enhance the dissolution process, and/or stabilize the amorphous state; the drug–surfactant ratio can also be used to affect the particle size.

Lyophilization

Lyophilization is a means of solvent or water removal resulting in a dry porous matrix, which can be easily reconstituted. Lyophilization is not considered by itself to be a particle formation technology, as often it is a dry cake which results. However, when coupled with some of the other technologies mentioned in this chapter, lyophilization provides a means of storage of particles formed by those technologies in a dry and compact state. Lyophilization essentially consists of a freezing step to temperatures below the eutectic (typically –40°C) in a vacuum, resulting in sublimation of the solvent. A primary and a secondary drying step complete the cycle. Whereas in supercritical fluid technology, one is interested in the part of the phase diagram where liquid and gas exist in a single state, freeze drying makes use of that part of the phase diagram where solid can be converted to gas bypassing the liquid state. Thus, both technologies utilize temperature and pressure, but at opposite extremes. Atmospheric (pressure) freeze drying, has been used by the Eskimos to effectively preserve meat. Freezer burn is another example of atmospheric freeze drying.

Self Assembly

Self assembly using amphiphilic components to trap drug in solution within a hydrophobic or hydrophilic core has been explored for decades for the delivery of drugs. The most notable example is that of the liposome, which, however, suffered from low entrapment efficiency, leakage, and rapid clearance by the reticuloendothelial system of the liver. This rapid clearance was reduced by the use of PEGylated phospholipids to form of the so-called sterically stabilized or stealth liposomes.

Self-assembly of phospholipids, such as phosphatidylcholine, a nutritional supplement, in the presence of calcium creates 50–500 nm large, continuous, solid, lipid bilayer sheets rolled into a spiral structure termed nanocochleates. These can be made to envelope a lipid-soluble drug, such as amphotericin B, to exclude water and protect against pH, oxidation, light, enzymatic attack, hydrolysis, and extremes in temperature. The nanocochleates will convert to liposomes at a pH greater than 6.5 in a calcium deficient environment as is found in the cell (typically less than 1–2 mM).

Recent efforts have focused on making novel high molecular weight polymers consisting, for example, of a linear chain of cyclodextrin alternating with PEG which self-assemble into uniform nanoparticles around the drug and protect it from enzymatic attack. Targeted delivery is achieved by derivatizing the polymer with cell surface receptor ligands. Because the molecular weight is greater than 50,000 kD, renal clearance is reduced, and plasma half-life is increased. pH-related chemistry within the cell causes the nano-particle to release its payload. The technology has been applied to camptothecin, as well as nucleic acids such as plasmid DNA or siRNA.

Challenges with Particle Engineering Approaches

While there are an abundance of particle engineering approaches, they are not a panacea, as each has its unique challenges. The main challenge facing these technologies is being able to produce drug product on a cost-effective manufacturing scale, and which meets the quality and safety standards of the regulatory agencies.

With nanosuspensions, one must be able to prevent agglomeration, or Ostwald ripening over the shelf-life of the product, as well as when in contact with biological fluids. If converted to the dry state, one must ensure redispersibility to the nano size over the shelf-life of the product. The disposition of ingested or injectable biodegradable and undissolved nanoparticles may need to be ascertained through appropriate toxicology testing. For processes such as bead milling and homogenization, the finished product should be essentially free of residual metal or grinding media.

Sterilization of particulate formulations intended for parenteral or pulmonary administration may require manufacture of sterile drug if terminal sterilization processes such as autoclaving or gamma irradiation result in chemical degradation, aggregation, or discoloration of the dosage form. Since sterile filtration involves pore sizes of 0.2 μm or less, this approach may not be feasible for nanosuspensions having a D_{50} of 200 nm or greater.

In the case of amorphous systems, absence of crystallinity upon tableting or over the shelf-life of the product must be ascertained to avoid potential slow down in dissolution, reduction in bioavailability, and loss of elegance. For lipid containing products, minimizing oxidation during manufacture and storage must be planned for by processing under inert conditions and incorporation of antioxidants in the formulation.

The spraying approaches require that the drug be soluble in either an aqueous or an organic solvent; many of the difficult to formulate drugs are neither. For those which are soluble in a solvent, the issue of residual solvent must be addressed, as well as that of environmental emissions control and operator exposure. Safety issues such as explosivity also need to be assessed for grinding techniques and those involving solvent. The safety of handling large quantities of small particles during manufacture also needs to be addressed from the standpoint of operator exposure.

17

Alfalfa in Molecular Farming

In 1986, it was shown that tobacco plants and sunflower calluses could express recombinant human growth hormone as a fusion protein. Since then, a diverse range of plant systems has been used for the production of pharmaceuticals. We have developed a production system based on the leaves of alfalfa (*Medicago sativa* L.), a choice made originally because of the plant's many favorable agronomic characteristics. Alfalfa is a perennial plant, so vegetative growth can be maintained for many years. For molecular farming, this characteristic, combined with the ease of clonal propagation through stem cutting, makes alfalfa a robust bioreactor with regard to batch-to-batch reproducibility. Among perennial plants, legume forage crops such as alfalfa have the advantage of fixing atmospheric nitrogen, thus reducing the need for fertilizers. Moreover, as a feed fodder crop, alfalfa has benefited from important research aiming to increase leaf protein content, so that today's varieties produce as much as 30 mg total protein per gram fresh weight.

In addition to these appealing agronomic characteristics, biotechnological research has revealed additional benefits for the production of pharmaceuticals in alfalfa. Expression cassettes have been optimized for protein expression in alfalfa leaves. Methods for transient protein expression have been developed so that it is now possible to use agroinfiltration or the transformation of protoplasts for early-stage demonstration and validation steps. In addition, glycosylation studies have shown that alfalfa is capable of producing recombinant glycoproteins with homogenous (uniform) glycosylation patterns.

This chapter provides an overview of the tools that have been developed and optimized specifically for the production of pharmaceuticals in alfalfa, with the emphasis on recent technological breakthroughs. The ability of alfalfa leaves to produce complex recombinant proteins of pharmaceutical interest is discussed and illustrated with recent data obtained in our laboratories. Data are presented concerning the production and characterization of alfalfa-derived C5-1, a diagnostic anti-human IgG developed by Hema-Quebec for phenotyping and cross matching red blood cells from donors and recipients in blood banks.

Alfalfa-specific Expression Cassettes

The first hurdle encountered during the development of alfalfa as a recombinant protein production system was the relative inefficiency of the available expression cassettes. A study in which a tomato proteinase inhibitor I transgene was expressed in tobacco and alfalfa under the control of the *cauliflower mosaic virus* (CaMV) 35S promoter showed that 3–4 times more protein accumulated in tobacco leaves compared to alfalfa leaves. Despite the low efficiency of the CaMV 35S promoter in alfalfa, biopharmaceutical production using this system has been reported in the scientific literature. Such reports include expression of the foot and mouth disease virus antigen, an enzyme to improve phosphorus utilization and the anti-human IgG C5-1. In this last work, the C5-1 antibody accumulated to 1 % total soluble protein.

Given the relatively high level of C5-1 antibody detected in alfalfa leaves using the weak CaMV 35S promoter, it was expected that expression cassette optimization would lead to significantly higher yields. The first family of expression cassettes we developed was thus designed to achieve strong expression in the aerial parts of alfalfa plants. The MED-2000 series (patent pending) consists of strong, leaf-specific expression cassettes, and is based on regulatory sequences from the alfalfa plastocyanin gene.

Using cassettes of this family to drive the *gusA* reporter gene, it was possible to achieve up to 14-fold the level of expression obtained in alfalfa leaves with the 35S promoter. Interestingly, although the MED-2000 promoters were derived from alfalfa genomic sequences, they also produced up to 25-fold higher β-glucuronidase (GUS) activity than the 35S-*gusA-nos* construct in the leaves of transgenic tobacco plants.

Because pharmaceuticals are bioactive molecules, their accumulation in plant cells could have a deleterious effect on the growth and development of the host plant. Therefore, we have developed a second series of expression cassettes incorporating inducible promoter elements. The regulatory elements of the MED-1000 series expression cassettes are derived from the alfalfa nitrite reductase (NiR) gene. The induction strategy used with these expression cassettes exploits the ability of alfalfa to grow abundantly in the absence of mineral nitrogen while fixing atmospheric nitrogen through its symbiosis with rhizobium, but also takes into account the fact that NiR genes are highly inducible by nitrate fertilizers.

We have demonstrated that the alfalfa NiR promoter is an excellent candidate for the inducible control of transgene expression in alfalfa leaves. As an example, a 3-kb genomic fragment corresponding to an alfalfa NiR promoter was isolated and fused to the *gusA* gene for analysis. We have shown that the promoter remains silent in nodulated plants grown in a nitrate-free medium. Upon the addition of nitrate, however, *gusA* gene expression is induced, and the reporter enzyme accumulates to a similar level to that observed in the leaves of 35S-*gusA* alfalfa plants.

Alfalfa Transformation Methods

Genetic transformation, which results in the stable integration of foreign DNA into the genome, is one of the key technologies underpinning the production of pharmaceuticals in alfalfa. Plant transformation at the industrial level must be optimized for efficiency, predictability and reproducibility in all aspects ranging from explant preparation to the physical conditions of DNA intake and the recovery of transgenic plants. This is an interesting challenge because plant transformation efficiency depends on many factors, including DNA conformation, explant type, plant species, plant genotype and the culture medium. In addition, the development of a plant-based expression platform to produce pharmaceuticals, nutraceuticals and industrial enzymes adds further requirements in terms of plant transformation. For example, a key issue in prototype development is the rapidity with which the ability of the system to produce a selected molecule can be tested, and this reflects the identification of optimal regulatory sequences to drive transgene expression. In order to address these various issues, we have adapted documented transformation methods and developed an alfalfa transformation portfolio ranging from proof-of-concept technology that allows rapid screening of target proteins, to stable expression in transgenic plants or cell cultures for sustainable commercial-scale production.

As for many plants, alfalfa is amenable to transformation by various methods including *Agrobacterium*-mediated transfer, direct DNA transfer to protoplasts using polyethylene glycol, and particle bombardment. In recent years, we have developed a medium-throughput system to manage the various activities related to plant transformation, from plant preparation through to transformation and regeneration. This allows us to maintain a continuous production schedule. The system allows the introduction of up to six constructs per week, which represents approximately 600 explants, and this capacity can easily be scaled up by increasing the number of staff and the availability of appropriate equipment. Thus far, more than 180 constructs have been integrated into alfalfa, and several thousand transgenic plants have been generated in our facilities. Given that 98% of the regenerated plants are PCR positive for the gene of interest, our medium-throughput system appears to work very efficiently.

In order to reduce the time required to confirm the accumulation of a given recombinant protein, we have developed a cell culture system in which transgenic alfalfa callus material produced at the proliferation step of *Agrobacterium*-based transformation is used to initiate cell cultures. These cell suspensions can be subcultured to sustain batch production of modest protein amounts. The protein blot demonstrates our ability to detect a recombinant protein in total soluble protein extracts from alfalfa cell cultures. It must be emphasized at this point that the recovered protein is most likely derived from several transformation events involving the same gene construct. This technique allows the detection of recombinant proteins 6–8 weeks after transformation, which is three times faster than the 20 weeks required to regenerate and screen transgenic plants following *Agrobacterium*-mediated transformation. This development has also shown that our alfalfa expression cassettes, although more adapted for leaf expression, provide adequate expression in cell cultures.

Although cell culture considerably reduces the time required to achieve proof-of-concept for new molecules, this time frame still needs to be reduced. In addition, there is some concern that the cell culture system might not correctly predict the ability of alfalfa to assemble complex proteins, and might not be a suitable guide for the selection of subcellular targeting strategies. We have therefore adapted several transient transformation methods to work with the alfalfa platform, including PEG- based protoplast transformation, particle bombardment and *Agrobacterium*-mediated transient transformation of leaves (agroinfiltration). The last method turned out to be particularly successful for the selection of optimal targeting strategies for a given candidate protein. More importantly, the relative protein accumulation in the different subcellular compartments is similar in leaves from agroinfiltrated and transgenic plants. In the case presented here, chloroplast targeting led to the highest accumulation both in agro-infiltrated leaves and transgenic plants, followed by targeting to the cytosol and mitochondria.

Agrobacterium-mediated transient gene expression has become the method of choice for rapid validation of gene constructs and targeting strategies in alfalfa leaves. In this system, an *Agrobacterium* culture carrying the T-DNA of interest is forced to enter into the intercellular spaces of the leaves under high vacuum. Once the physical barrier of the epidermis is crossed, the bacteria infect neighboring cells, transferring T-DNA copies into the nucleus. Although the T-DNA exists inside the nucleus only transiently, the genes present on the T-DNA are transcribed, leading to the production of the recombinant protein in each infected cell. The efficiency of this method is thus highly dependent on the ability to distribute the bacterial culture evenly inside the leaf tissue.

As well as its short time frame, agroinfiltration has several further advantages for recombinant protein production. The method allows the expression of multiple genes by infiltrating cells with a mixture of two or more *Agrobacterium* cultures (co-infiltration), thus eliminating the need to clone several genes within the same T-DNA. Agro-infiltration is also readily scalable. Routinely, 25 leaves are infiltrated for immunological verification of expression or the comparison of targeting strategies. However, after the selection of an ideal transgene construct, infiltration of 7500 leaves per week can be carried out by a limited number of staff, in a continuous process, for the production of micrograms of recombinant protein.

The production of C5-1 by co-infiltration illustrates the impressive capacity of this method. Results show that the production of C5-1 in detached alfalfa leaves was validated within 5 days from infiltration. In these experiments, different bacteria bearing the light- and the heavy-chain constructs were used to infect the cells. Most of the infected cells were occupied by both strains, and a protein corresponding to fully assembled C5-1 was detected in the infiltrated leaf extract. This result demonstrates the potential of agroinfiltration for testing the adequate expression and assembly of complex proteins in alfalfa leaves using different *Agrobacterium* strains.

Characteristics of Alfalfa-derived Pharmaceuticals

When recombinant proteins are produced in a heterologous system, there may potentially be differences between the final product and the natural molecule. Hence, for each new protein produced in alfalfa, a thorough analysis of the processing, folding, assembly and post-translational modification is conducted to ensure the conformity of the purified molecules. This section describes the analysis of alfalfa-derived C5-1 antibodies to demonstrate the ability of alfalfa plants to produce large amounts of high-quality molecules for therapeutic or diagnostic applications.

Purified C5-1 has been obtained from alfalfa leaf extracts by affinity chromatography on either a human IgG-Sepharose column or a Streamline rProtein A-Sepharose column. Interestingly, the purified product obtained with these two methods differed significantly. The antibody fraction obtained from the human IgG column contained a mixture of different intermediate assembly forms of the heavy (H) and light (L) chains, ranging from H2 to the fully assembled H2L2 form. In comparison, purification on rProtein A-Sepharose resulted in the isolation of H2L2 form alone. This situation emphasizes the major impact that purification methods can have on the characteristics of the end product.

In some heterologous production systems, improper removal of the signal peptide may occur during the expression of secreted proteins, which would result in the addition or removal of amino acids at the N-terminal end. In most cases, these modifications are undesirable in a therapeutic context. For C5-1 expression in alfalfa, the natural sequence encoding the signal peptide was retained during the assembly of the expression cassettes. Although most examples show that mammalian signal peptides are correctly processed in plants, N-terminal amino acid sequencing was performed on the heavy chain of alfalfa-derived C5-1 in order to confirm the N-terminal integrity of the antibody. The N-terminal sequence of the heavy chain was confirmed as EIQLV, which is identical to that of the hybridoma-derived C5-1 and indicates the correct processing of the signal peptide in alfalfa.

N-glycosylation is another important issue when considering the conformity of therapeutic proteins produced in heterologous systems. Although every eukaryotic expression system N-glycosylates proteins targeted to the secretory pathway, each system links a different form of N-glycan to the recombinant protein. The glycans synthesized in a heterologous production system only rarely correspond to those found

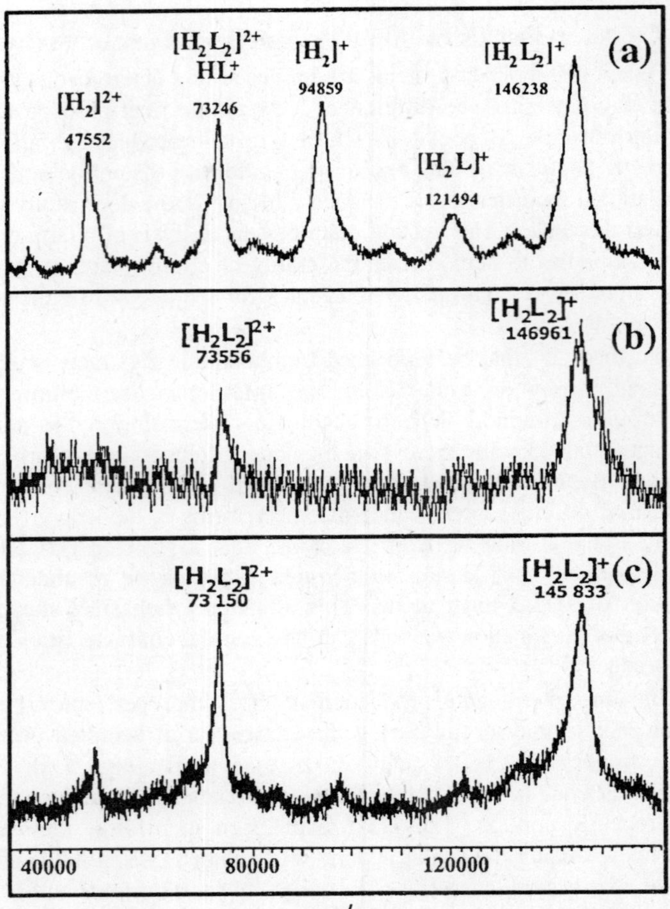

Fig. 17.1. MALDI-TOF mass spectra of purified alfalfa-derived C5-1 using (a) human IgG or (b) protein A. (c) Hybridoma-derived C5-1.

in the natural source of the protein. In this context, the ability of plants to perform complex glycosylation represents an advantage over yeast and insect cells, and places the plant system in the group of Chinese hamster ovary cells (CHO) and murine myeloma cell lines (NSO). Importantly, however, the analysis of recombinant IgGs produced in tobacco indicates heterogeneity in the structure of N-glycans.

In contrast, glycosylation analysis of alfalfa-derived C5-1 showed that a single, unique N-linked glycan form is found on the antibody. The glycoform is representative of plant complex N-glycans, and includes

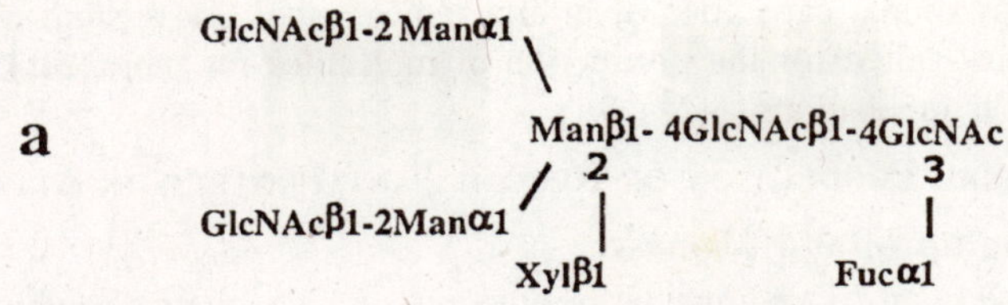

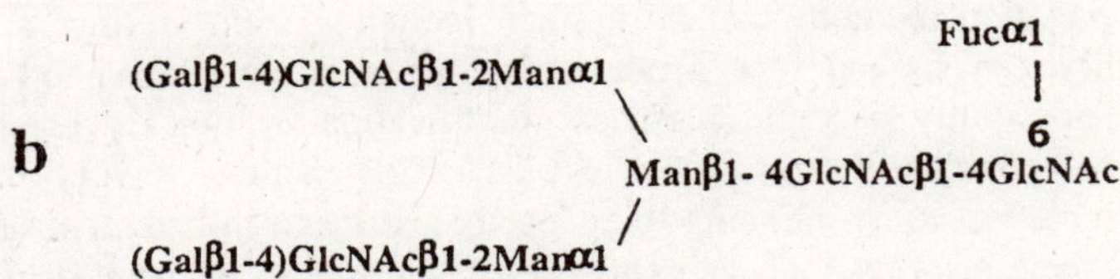

Fig. 17.2. Structure of N-glycans from (a) alfalfa-derived C5-1 and (b) murine C5-1.

core β(1,2)-xylose and β(1,3)-fucose. Homogenous N-glycosylation of a recombinant protein ensures batch-to-batch reproducibility, but also provides an ideal substrate for in vitro modification of the N-glycan. For example, it has been shown that incubating the purified alfalfa-derived C5-1 with β(1,4)-galactosyltransferase in the presence of UDP-galactose resulted in an efficient addition of β(1,4)-galactose to the terminal GlcNAc residues of the N-linked glycans.

Table 17.1. Activity of alfalfa- and hybridoma-derived C5-1.

Extract	*Specific activity (OD/100 ng)*	*True affinity (K_Ds)*
C5-1 from hybridoma	0.2 35 ± 0,020	4.6×10^{-10} M
C5-1 from alfalfa	0.267 ± 0,080	4.7×10^{-10} M

In order to compare the specific activity of plant-derived C5-1 to that of the hybridoma-derived antibody, the antigen-binding capacity of antibodies produced in each system was assayed by enzyme-linked immunosorbent assay (ELISA). As shown in **Table** 1.1, antibodies from both sources demonstrated similar binding characteristics against human IgGs. Furthermore, the stability of alfalfa-derived C5-1 in the blood stream of Balb/c mice was comparable to that of the hybridoma-derived IgG.

In the light of the results presented above, we conclude that alfalfa offers a suitable system for the high-yield production of correctly assembled complex proteins, including multimeric glycoproteins. The post-translational capacities of alfalfa indicate that this system is one of the best-suited for the production of molecules for therapeutic and diagnostic applications.

Industrial Production of Recombinant Proteins in Alfalfa

Ramping up Alfalfa Biomass

In addition to yielding large amounts of high quality protein, an efficient system for large-scale recombinant protein production should include a rapid biomass amplification method. In alfalfa, different propagation methods can be applied including stem cutting, somatic embryogenesis and seed production. The stem cutting method offers the possibility of rapid biomass amplification by quickly creating a clonal population from an elite plant. This method is very reliable with respect to maintaining the capacity and batch-to-batch consistency of crucial aspects such as expression level and product uniformity.

Alfalfa stem propagation can be achieved without the addition of hormones so long as the cuttings are maintained in a humid environment. One transgenic alfalfa plant can generate a clonal population filling a 1000-m^2 greenhouse within 14 months. Stem cutting is also the method of choice when small quantities (milligrams) of recombinant protein are required within a limited time frame, as it is the case for product testing and pre-clinical studies.

Alfalfa Harvest, and Recovery of Recombinant Molecules

In greenhouses, alfalfa can be harvested 8–10 times per year using minimal equipment. A 5-ha greenhouse containing mature alfalfa plants will yield about 130 tons of fresh biomass at each harvest, if 75% of the surface is cultivated. Harvested 10 times, such a greenhouse will generate 900 tons of alfalfa annually, although for practical reasons the harvests are distributed throughout the year. For example, within the same greenhouse, the plants are distributed into 10 plots of 0.375 ha each. If each plot is harvested every 5 weeks, this means that 100 harvests of 13 tons are performed per year. This biomass of fresh alfalfa tissue can easily be handled by medium-scale processing machinery.

Extraction of soluble proteins from alfalfa tissue begins with a maceration step in a hammer mill. This step is intended to break the cells to facilitate extraction of confined water and soluble content. From the resulting mash, the green juice is extracted using a screw press. Four presses, each with a 400 kg h^{-1} biomass intake capacity will produce 800 L h^{-1}, making a total of 6500 L of green juice per harvest. This capacity is essential to minimize the delay between harvest and extraction. The soluble protein content of alfalfa green juice produced in this manner is about 2%, so a 5-ha greenhouse generates 130 kg of soluble proteins twice weekly. Finally, with current recombinant protein expression levels of 0.1% to 1% of soluble proteins, the yield from the 5-ha greenhouse is estimated to be between 13 kg and 130 kg of recombinant protein per year.

Purification of recombinant proteins from alfalfa extracts can be performed using several strategies, depending on the required purity and the intended application of the protein. Sophisticated and powerful methods are under development in our laboratories. In the case of C5-1, two purification methods have been investigated. The first method is an adaptation of the method used at Hema-Quebec for the purification of C5-1 from hybridoma cells. It involves affinity chromatography using a human IgG1 (the antigen recognized by C5-1) to purify the protein. Alternatively, an expanded bed affinity chromatography method using staphylococcal protein A has better potential for larger scale preparation of C5-1. These large-capacity columns, in which the liquid flows upward, can be loaded with unclarified green juice, and can bind up to 20 mg of human IgG per mL of medium. Unfortunately, protein A has a low affinity for mouse IgGs. In our hands, at pH9, we have obtained up to 5 mg C5-1 per mL of medium. Current improvements of the purification step will include coupling the expanded bed column with streptococcal protein G, which shows a higher affinity for mouse IgGs than protein A.

Although greenhouses can supply enough biomass to produce kilograms of recombinant proteins, field production would be necessary if tons of recombinant proteins were required. At this scale, molecular farming will benefit from the current knowledge developed for the animal feed industry. Among the most significant developments impacting on large-scale alfalfa processing, the wet fractionation process, currently used by Sativa 2000 in Champagne (France), treats up to 750,000 tons of fresh alfalfa per year. In the wet fractionation process, temperature and pH are used to separate alfalfa proteins and isolate protein-rich fractions, which are dried into pellets for the animal feed industry. A refined version of this large-scale protein separation process is used at Viridis to produce purified Rubisco from alfalfa as a food additive for human consumption.

The intrinsic qualities of alfalfa justify its selection as a platform for the production of heterologous proteins. Alfalfa plants are easily propagated by stem cutting to create large populations. In greenhouses, these populations can be harvested 10 times per year, and the plants can be maintained for more than 5 years. Alfalfa is also capable of producing and processing complex proteins, and adds homogenous N-glycan chains to secreted glycoproteins.

The expression cassettes developed in our laboratories facilitate the strong expression of recombinant proteins in alfalfa leaves. High levels of the candidate molecules accumulate when they are targeted to the most appropriate subcellular compartment. A rapid capacity to evaluate expression strategies has been developed based on the accumulation of proteins in agroinfiltrated leaves. By combining this rapid selection method with the efficient genetic transformation of alfalfa, the solid foundation of an effective heterologous expression platform has been secured.

INDEX